LASER-INDUCED MATERIALS AND PROCESSES FOR RAPID PROTOTYPING

LASER-INDUCED MATERIALS AND PROCESSES FOR RAPID PROTOTYPING

by

L. Lü
J. Y. H. Fuh
Y.S. Wong
The National University of Singapore

SPRINGER SCIENCE+BUSINESS MEDIA, LLC

Library of Congress Cataloging-in-Publication Data

Lü, L. (Li), 1954 –
Laser-induced materials and processes for rapid prototyping / L. Lü , J.Y.H. Fuh, Y.S. Wong
p. cm.
Includes bibliographical references and index.
ISBN 978-1-4613-5569-4 ISBN 978-1-4615-1469-5 (eBook)
DOI 10.1007/978-1-4615-1469-5
1. Lasers—Industrial applications. 2. Prototypes, Engineering. I Fuh, J.Y.H.II. Wong, Y. S. (Yoke-San) III. Title.

TA1677.L82 2001
621.36`6—dc21

2001035423

Originally published by Kluwer Academic Publishers in 2001
Softcover reprint of the hardcover 1st edition 2001

Printed on acid-free paper.

TABLE OF CONTENTS

PREFACE

The term rapid prototyping (RP) refers to a generic group of emerging technologies that enable very quick fabrication of engineering components primarily targeted for prototyping applications. With RP, very complex three-dimensional parts or prototypes can be fabricated without the need of costly tooling and machining. This inevitably leads to much shorter design cycle time and lower cost of building a prototype. Its manifold benefits include significant productivity gains, cost saving, and shortened development time to introduce concept models. As such, RP technologies have attracted tremendous R&D interests from both academia and industry in the past decade. Many different processes and materials have been commercialized and used in industry primarily for the fabrication of physical prototypes. More recent interests in RP technologies are towards functional applications of the fabricated parts, such as in rapid tooling applications and replacements of damaged components. Many processes and materials have been commercialized but are yet to be able to fulfill the aforementioned functional requirements because of limited mechanical strengths of the fabricated parts. One major problem lies in the materials used in these processes. Research on new material compositions and characterization, coupled with appropriate design and control of the associated process parameters, is expected to play an increasingly crucial role in RP development, particularly in relation to the direct fabrication of functional parts for engineering applications

In this book, the authors focus on some of the fundamental issues on RP materials used in laser-induced processes. In particular, the laser sintering of metal powder materials is thoroughly examined in addition to a detailed discussion on the most widely used polymer-based laser-lithography process. The commonly observed problems encountered when applying RP techniques are discussed based on the recently published research works. The remedies and possible solutions to issues pertaining to materials and processes are included. The book contains nine chapters. Chapter 1 presents an overview and background on RP technologies and future R&D direction. The material and process issues are highlighted as these are the main contents of this book. Chapter 2 details the fundamentals of the laser-lithography, also known as StereoLithography Apparatus (SLA) process. The principles and physics of laser scanning and the formation of multi-scanned lines and multi-layer models are presented. The characteristics of photo-polymerization are also discussed in depth. The mechanics and curing behaviors from laser-induced polymerization are examined in Chapter 3 with experimental results. Findings from differential scanning calorimeter and Moiré analysis method have been used to improve the current SLA systems significantly. Functional enhancements through the use of fibers and AEROSIL powders as reinforcement in photopolymer are presented in Chapter 4.

Chapter 5 describes the mechanisms of the laser sintering processes and material characterizations. The principle of infiltration process and densification of the laser-sintered parts are introduced. Due to the existence of porosities using laser sintering, a technique based on direct laser melting is introduced in Chapter 6, in which the formation of several common metal systems, including Cu-, Ni- and Fe-based compounds, is discussed. The major advantage of direct laser melting is that the powder is fully melted and generally no porosity exists. Therefore, no post-processing is required. Besides metal systems, the formation of composites, such as TiC, TiB_2, etc. using ex-situ and in-situ processes, is described in details due to their potentials for wider usages in functional tooling. The ceramic systems applicable to RP systems are introduced in Chapter 7.

To select or improve an RP material for a specific RP process, it is necessary to understand the basic characteristics and mechanisms of different RP techniques. Chapter 8 focuses on process characterization and modeling, as well as optimization approaches applied to RP systems, including important topics such as adaptive slicing and optimal orientation. These approaches are applicable to most RP processes to improve their effectiveness and efficiency. Chapter 9 describes the rapid tooling technology with a detailed case study on an injection mold tooling produced using the direct metal laser sintering method. Examples include building injection tooling (cavity and core inserts) via RP and the development of EDM electrodes based on potential new materials are described. It aims to highlight some of industrial applications of rapid tooling based on the existing RP techniques.

This book targets those who want to know laser-based RP technologies in greater details. Most of the reported works are based on the research results (technical reports, papers and patents) from the authors in the past seven years at The National University of Singapore through cross-disciplinary research spanning manufacturing, materials and control areas. It emphasizes material issues whilst also addresses the main RP processes. So far, there are very few books that address the material development and the associated fundamental behaviors of RP processes; this book hopes to fill part of the gap and aims to serve as a comprehensive reference to those who want to know and understand more about laser-based RP technologies from the materials viewpoint. It can be used as graduate-level textbook materials or a reference book for those engineers and scientists in industry who are working on RP research and development. The authors believe that the issues highlighted and results presented are not only informative, but will also very useful for future commercial developments.

L. Lü
J.Y.H. Fuh
Y.S. Wong

ACKNOWLEDGEMENTS

The authors would like to thank the following individuals who have made this book possible:

- Professor Andrew Y.C. Nee who has co-supervised many research students, provided crucial guidance and contributed to the results published from the group;
- Associate Professors Han Tong Loh, Yoo Sang Choo and Zexiang Shen who have co-advised research students and staff in RP research, and Professor En Tang Kang for his invaluable advice on polymer analysis;
- Professors Beno Benhabib, Chull B. Park and Gen Zak, our overseas research collaborators from University of Toronto in one key RP project funded by the Singapore-Ontario Joint Research Program;
- Manlong Sun, Chi Mun Cheah, Tatso Miyazawa, Fen Xu, Yaxin Tang, Xinhua Wang, Wanlong Wang, Jianguo Wang, Yong Soon Lim, Yeow Hui Tan, Hlwan Moe Zaw, Min Wai Khaing, Joo Khai Toh, Kam Choong Lee, Shor Hong Ho, and Cheng Cheh Tan for contributing their research findings;
- Department of Mechanical Engineering and Faculty of Engineering at the National University of Singapore for providing various RP research grants and resources;
- Gintic Institute of Manufacturing Technology in collaborating, Productivity and Standard Board of Singapore in providing laser processing equipment, and the RP group from Nanyang Polytechnic for collaborating and making available their RP systems and injection molding facilities for case studies; and

- National Science and Technology Board of Singapore for funding, SONY Precision Engineering Center, Singapore and SONY Corporation, Japan for collaborating and sponsoring the RP research, and Kinergy Pte. Ltd. for contributing to some of the recent works.

Finally, we would like to express a special gratitude to our families for supporting us in the lengthy preparation of this publication. Their understandings and supports have enabled us to complete this book in time.

L. Lü
J.Y.H. Fuh
Y.S. Wong

Chapter 1

INTRODUCTION

1.1 Overview of rapid prototyping and manufacturing

Rapid prototyping and manufacturing (*RP&M*) [1] or simply RP as used in this book, and also called *layered manufacturing*, *solid free-form fabrication*, and *3D printing*, refers to a generic group of emerging technologies that enable quick fabrication of engineering components primarily targeted for prototyping applications. It is an important and significant technological break-through in the manufacturing industry [2]. Compared to the conventional material forming and removal processes, the part is produced by an additive process through gradual creation or addition of solid material to the required shape [3,4]. This advanced manufacturing technology, which combines laser technology, photo-chemistry and computer graphics, is capable of direct and rapid manufacture of complex three-dimensional objects from profiles created using CAD, X-ray, Computer Tomography (CT)/Magnetic Resonance Imaging (MRI) scanning, and other data. This capability, coupled with the fact that no fixtures and tooling are required in the process, has brought about a very significant reduction in the cost and lead time taken to introduce a new product to the market.

The technology emerged in 1987 with the introduction of first commercialized RP machine called StereoLithography Apparatus (SLA) from 3D systems. It has now been applied to many engineering fields, e.g. aerospace, automobile, medical, consumer products, business machines, etc. The potential for incorporating RP with science and art has also been recently reported [5]. More than 30 processes have now been developed world-wide and more than 22 RP system manufacturers market the technology. Although there is a declining demand in RP machines from industry since 1998 due to the growing number of RP service bureaus (SBs), 4,259 systems have been sold to industry, academic, and government sites around the world at the end of 1999 [6]. The estimated worldwide revenues for RP products and services reached US$520M in 2000 [6]. More than 334 SB locations have been established according to a recent report [6,7]. Further details of RP development, history and technology can be found in References [1, 6, 8-11].

Although called "rapid" prototyping and manufacturing, the actual layer-by-layer fabrication process can be relatively slow and time-consuming when compared to the conventional manufacturing processes. The main benefit of rapid manufacturing is the saving in time and cost on tooling and re-engineering. RP allows not only quick fabrication of models for design verification, but also functional components and tooling. The tooling produced via RP techniques is often termed rapid tooling (RT). In this book, we will refer RT to be a special case of rapid manufacturing. With the wider adoption of indirect and direct tooling techniques in industry, the shift from normal prototyping practice to RT is underway [12] with increasing interests in extending RP technologies to RT.

1.2 Laser-induced rapid prototyping

Among the common and matured RP processes, those based on the use of a laser for materials processing are the Stereolithography (SL) process, the Laminated Object Manufacturing (LOM), and the Selective Laser Sintering (SLS) process. The basic building mechanisms of the SL and SLS processes, which are the most important laser-induced RP processes, will be presented in details in Chapters 2 to 5. Some of the fundamental issues on RP materials used in the aforementioned laser-induced processes, particularly the laser sintering of metal powder materials and polymer-based stereo-lithography process, will be discussed in the subsequent chapters as the main focus of this book.

1.2.1 Laser-lithography (LL) process

The LL process (as called in this book) creates three-dimensional parts (see Fig.1.1) by selectively solidifying polymeric materials layer-by-layer upon exposure to UV radiation or laser beams. It is still the most accurate RP process in terms of dimensional accuracy and capability in creating small fine features. Although patented in 1986 by 3D Systems, LL has continued to grow from its original application of producing concept models rapidly for visualization and verification towards manufacturing. Rapid prototypes produced using the LL method have already been successfully used for form-fitting functions and as patterns for the production of molds and castings. Research and development works have also been carried out for the production of EDM tools [13,14] from LL parts. Despite its great application potential, prototype parts created by currently available LL systems exhibit weak mechanical properties and significant amount of shrinkage. These inherent weaknesses hinder its wider adoption, especially for applications involving mechanical loading. Hence, enhancement of the functionality of LL-built part is being sought for higher accuracy of the built parts and improved material properties of the SL resins.

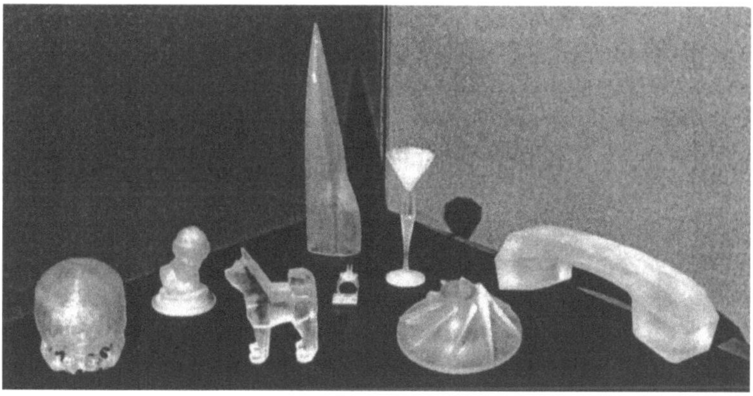

Fig.1.1 3D parts created by LL process.

Several major drawbacks in the LL process are still imminent, mainly from the inherent physical and chemical properties of LL resins and the required trial-and-error procedures in fine-tuning the process parameters. From the material point of view, major problems are in the significant amount of shrinkage encountered during photo-polymerization and its weak mechanical properties. Distortions and warping of a part due to shrinkage, and internal stress build-up during the layered manufacturing process greatly influence the dimensional accuracy of the part being built [15]. These distortions in the laser-scanned part are significantly increased during the post-curing process. The distortion and the weak mechanical properties limit the functionality of the part. Therefore, Chapters 2 to 4 focus on the fundamentals of the LL process control and material study to provide a good basis for research to improve LL-built part quality.

1.2.2 Selective laser sintering process

The selective laser sintering (SLS) process was developed at the University of Texas at Austin by Beaman and Deckard [16, 17]. In the SLS process, a layer of powder is deposited on a support and leveled by a rolling device. Similar to LL process, a laser beam scans a two-dimensional pattern on the deposited powder layer to sinter the powder layer. After sintering of a layer, a new layer of powder is deposited in the same manner. By successive powder deposition and laser scanning, a 3-D part is then built [18, 19].

In the first three years of commercial use, the SLS process was used primarily to produce durable nylon-based prototype and patterns for making silicone rubber molds. Rubber molds are used to vacuum cast durable plastic prototype parts from liquid urethane or epoxy systems. The direct production of nylon-based parts is generally economical when a limited number of parts is needed. The vacuum casting is attractive when more parts are required, particularly if the parts are larger. While nylon-based and cast prototype parts are useful for some level of functional evaluation, it is necessary to have prototype parts made with the production material and process if more rigorous testing and analysis are required. As a result, there has always been a great deal of interests in using rapid prototyping processes to produce durable prototype molds for injection molding plastic parts. In all of these examples, a rapid prototyping process is used to make a plastic pattern that is then converted to a mold using a secondary process. While these options have some attraction, particularly as the quality of rapid prototyping patterns has improved, they all suffer from limitations associated with fabrication time, accuracy, durability, or a combination of these factors. Therefore, the SLS process has since been developed to fabricate metallic [20-25], plastic (see Fig. 1.2), ceramic [26-41] and composite [42] functional parts.

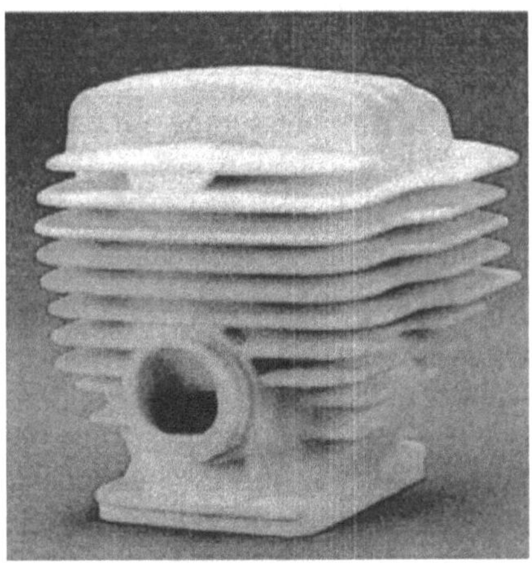

Fig.1.2 A polycarbonate pattern built by SLS process for investment casting [12].

There are basically two types of SLS processes: indirect and direct SLS processes. In the indirect SLS process, low temperature polymers, such as polymethyl methacrylate and copolymer, are used as binders. The polymer binders are either mixed with primary powders at a high melting temperature or coated on the surface of the primary powders [43,44]. DTM at Austin commercialized an indirect SLS process [45] using steel powder coated with polymer of 5 μm thickness. The laser beam melts only the binder, binding the primary powder particles together to form green parts. The strength of the green part is about 3 MPa. For it to be functional, the green part typically has to be post-heat treated, such as the infiltration process. The advantage of indirect SLS is the low laser power used. The main disadvantage is the tedious post-treatment.

To overcome the disadvantage of indirect SLS, direct SLS was developed. The main difference of direct SLS from indirect SLS is the use of the metal powder as a binder. Early investigation of direct SLS used single-phase metals, such as lead, zinc and tin. It was not successful due to balling phenomenon, where spherical globules were formed during solidification of the laser melted material [46]. In an attempt to mollify the balling tendency, a two-phase powder approach was developed [47]. In this approach, the material system must consist of at least two different powders with significant difference in the melting temperature. The direct SLS involves directly melting and consolidating selected regions of a powder bed to form a desired shape having high or full density.

EOS at Germany commercialized a direct SLS process called Direct Metal Laser Sintering (DMLS) using a bronze-nickel powder mixture with a small amount of copper phosphorus (Cu-P). When the lower melting metal is melted, such as Cu-P in bronze-nickel system, it infiltrates into the primary powder skeleton under capillary force [23]. EOS has claimed that parts fabricated using their newly developed powder have very low shrinkage.

In principle, only the powder particles or coatings with low melting temperature are melted in the traditional direct SLS process. The chemical composition is not homogeneous after SLS. Sintering or even HIPing process is still needed to obtain a uniform and densed structure. By increasing the laser energy density, the metal powder can be fully melted to form solid parts. The advantage of this selective laser melting (SLM) process is that no post process is required. However, since the powder is fully melted, the viscosity of the liquid phase must be well controlled. Accuracy of parts obtained with SLM is in general poorer than that achievable with SLS.

Based on the principle of laser cladding, AeroMet Corporation of Eden Praire, Minnesota developed the Lasform^{SM} process [48-50]. The development of this process began more than 20 years ago and technical feasibility was demonstrated as part of an R&D program funded by the U.S. Defence Advanced Research Projects Agency, with the Office of Naval Research Service as the technical agent [51]. Fully densed parts with no shrinkage can be fabricated using this selective laser cladding (SLC) process [51-59].

1.3 RP process characterization and modeling

The introduction of rapid prototyping technology in recent years has reduced the time to make a prototype from a CAD design and shortened the product development process. However, the application and usage of RP facilities are presently not well integrated into the entire product development cycle. This reduces the effectiveness of RP in the evolution of the product development process. A major factor hindering the integration is the lack of a comprehensive database or knowledge base to provide decision support for the selection of machine/process system. In addition, to select or be able to improve a RP process/material system, it is important to first characterize and model the building mechanisms of different RP processes.

Comprehensive benchmark studies have been carried out to assess achievable dimensional accuracy and surface quality of current RP processes [60, 61]. Such studies can provide useful decision support data for prototyping and product development. As RP processes are essentially based on the principle of successive layer creation and stacking or building, generic models can be developed that are applicable across several types of RP processes. From the experimental study and theoretical analysis, models for the computation of manufacturing time/manufacturing cost and the evaluation of surface quality/geometrical errors have been proposed and developed [62]. Process planning issues, such as material/machine selection, orientation determination, adaptive layer thickness slicing, have also been identified and examined. An appropriate RP material/machine selector can be developed based on an understanding of the RP process model and process planning issues. By incorporating a comprehensive database and a process-oriented knowledge base, the system is expected to provide full support for the transformation from a CAD design to a satisfactory RP part, and speed up the product development cycle. It also facilitates the development of suitable material systems to meet specific RP requirements.

1.4 RP technology trends

RP will inevitably continue to streamline the product development cycle. Although it is difficult to predict the impact of RP and RT technologies, we can nevertheless deduce some future directions based on the technology trends. The growing number of new processes and materials indicates the increasing potential and application of RP in more areas. The demand of using RP in medical application has been strong since the late '90. New applications include architecture, bioengineering, virtual prototyping, micro-mechanical-electrical-systems (MEMS), etc., and have widen the adoption of RP technologies by industry.

Although LL, SLS and DMLS will continue to meet the need of most RP users if good accuracy, strength or resistance is the main concern, cheaper and/or faster machines will increase their demands. Many compact and lower-cost desktop RP machines have been coming into the market recently. Those with more diversified multi-functional materials to cater for a wider range of applications will have a lead over other processes. Although indirect tooling currently has greater commercial success in the past five years, direct tooling is expected to have a better edge in the longer term (six to 10 years) [63]. There are still many issues on materials, process optimization and supporting software that need to be further researched and developed to bring the RP and RT technologies to further maturity.

1.5 References

1. P.F. Jacobs, *RP. & Manuf., Fundamentals of StereoLithography*, SME, Dearborn, MI (1992). 1.
2. D. Kochan, C.K. Chua, Z.H. Du, *Comp. in Industry*, Vol39 (1999), 3.
3. J.P. Kruth, *Annals of CIRP*, Vol. 40 (1991), 603.
4. J.P. Kruth, *Annals of CIRP*, Vol 47 (1998), 525.
5. M. Rees, *RP. J.*, Vol. 5(4) (1999), 154.
6. T.T. Wohlers, *RP. & Tooling, State of the Industry: 1999 Worldwide Prgress Report*, Wohlers Associates, Inc., Fort Collins, CO. (1999), 25.
7. *RP Directory* 1999 Edition, CAD/CAM Publishing Co. (1999), San Diego.
8. P.F. Jacobs, *Stereolithography and Other RP&M Technologies*, SME (1996), Dearborn, MI.
9. J.J. Beaman, J.W. Barlow, *et al. Solid Freeform Fabr.: A New Direction in Manufacturing*, Kluwer Academic Publishers, Norwell, MA (1997).
10. C.K. Chua, K.F. Leong, *RP. Principles and App. in Manuf.*, John Wiley & Sons (1997), Singapore.
11. Y.G. Wang, *RP. Techn.*, Huazhong University of Science & Technology Press (1999), Wuhan, China, (in Chinese).
12. A. Rosochowski, A. Matuszak, *J. Mater. Proc. Techn.*, 106 (2000), 191.
13. A. Arthur, P. Dickens, C. Bocking, R. Cobb, *Proc. Solid Freeform Fabr. Symp.*, Ed. D. Bourell, J.J. Beaman, H.L. Marcus, R.H. Crawford and J.W. Barlow, The University of Texas, Austin, Texas, 12-14 Aug. 1996,175.
14 R. Ippolito, L. Iuliano, A. Gatto, *Proc. Solid Freeform Fabr. Symp.*, Ed. D. Bourell, J.J. Beaman, H.L. Marcus, R.H. Crawford, and J.W. Barlow, The University of Texas, Austin, Texas, 12-14 Aug. 1996, 199.
15. Iwanaga, S., Ohkawa, *Proc. of 3rd Inter. Confer. on RP.*, Dayton, Ohio, (1992), 77.
16. J.J. Beaman and C.R. Deckard, *Selecive Laser Sintering with Assisted Power Handling*, US patent No.4,938,816, June 1990.
17. D. Girouad, *Proc. of the IBEC'93*, Detroit, 10.
18. G. Zong, Y. Wu, N. Tran, I. Lee, D.L. Bourell and J.J. Beaman, *Proc. Solid Freeform Fabr. Symp.*, Ed. H.L. Marcus, J.J. Beaman, J.W. Barlow, D.L. Bourell and R.H. Crawdford, The University of Texas at Austin, Austin, TX, 3-5 Aug. 1992, 72.
19. D.L. Bourell, R.H. Crawford, H.L. Marcus, J.J. Beaman and J.W. Barlow, *Manuf. Sci. Eng.*, ASME, Vol. 68 (1994), 519.
20. M. Agarwala et al., *RP. J.*, Vol. 1, (1) (1995), 26.
21. C.R. Deckard, *U.S. Patent* 4,863,538.
22. H.L. Marcus, J.J. Beaman, J.W. Barlow and D.L. Boarell, *JOM*, Vo. 42, (4) (1990), 8.
23. A. Lohner, *Metal Powder Report*, Vol. 52, No.2 (1997), 24.
24. M.K. Agarwala, D.L. Bourell, B. Wu, and J.J. Beaman, *Proc. Solid Freeform Fabr. Symp.*, Ed. H.L. Marcus, J.J. Beaman, J.W. Barlow, D.L. Bourell and R.H. Crawford, The University of Texas, Austin, Texas, 9-11 Aug. 1993, 193.
25. M. Agarwala, D. Bourell, J. Beaman, H. Marcus and J. Barlow, *RP. J.*, Vol. 1, (2) (1995), 36.
26. U. Lakshminarayan and H.L. Marcus, *Proc. Solid Freeform Fabr. Symp.*, Ed. H.L. Marcus, J.J. Beaman, J.W. Barlow, D.L. Bourell and R.H. Crawdford, The University of Texas at Austin, Austin, TX, 3-5 Aug. 1992, 44.
27. G.A. Brady and J.W. Halloran, *RP. J.*, Vol.3 (1997), 61.
28. E.A. Griffin, D.R. Mumm, D.B. Marshall, *Amer. Ceram. Soc. Bull.*, Vol. 75, No. 7 (1996), 65.
29. M.K. Agarwala, V.R. Jamalabad, N.A. Langrana, A. Safari, P.J. Whalen and S.C. Danforth, *RP. J.*, Vol. 2 (1996), 4.
30. V. F. Janas, and A. Safari, *J. Amer. Ceram. Soc.*, Vol. 78, (11) (1995), 2945.

31. K. A. Klicker, J. V. Biggers, and R. E. Newnham, *J. Amer. Ceram. Soc.*, Vol. 64, No. 1 (1981), 5.
32. J.D. Cawley, A.H. Heuer, W.S. Newman and B.B. Mathewson, *Amer. Cer. Soc. Bull.*, Vol. 75, (5) (1996), 75.
33. M. Greulich, M. Greul and T. Pintat, *RP. J.*, Vol. 1, (1) (1995), 20.
34. H.L. Marcus, J.J. Beaman, J.W. Barlow and D.L. Bourell, *Amer. Ceram. Soc. Bull.*, Vol. 69, (6) (1990), 16.
35. M.K. Agarwala, D.L. Bourell, A. Manthiram, B.R. Birmingham and H.L. Marcus, *J. Mater. Sci.*, Vol. 30 (1995), 459.
36. P. K. Subramanian, N. Vail, J. Barlow and H. Marcus, *RP J.*, Vol. 1, No. 2 (1995), 24.
37. H.H. Bache, *J. Amer. Ceram. Soc.*, Vol. 53, (12) (1970), 654.
38. I. Lee, *J. Mater. Sci. Lett.*, Vol. 17 (1998), 1907.
39. U. Lakshminarayan and H.L. Marcus, *Mater. Manuf. Proc.* Vol. 9, (5) (1994), 921.
40. P.K. Subramanian and H.L. Marcus, *Mater. Manuf. Proc.*, Vol. 10, (4) (1995), 689.
41. A. Manthiram, D.L. Bourell and H.L. Marcus, *JOM*, Nov. 1993, 66.
42. L. Lu, J.Y.H. Fuh, Y.S. Wong and C.C. Lim, *Mater. Res. Bull.*,Vol. 30 (2000), 1555.
43. N.K. Vail, J.W. Barlow, J.J. Beaman, H.L. Marcus and D.L. Bourell, *J. Appl. Poly. Sci.*, Vol. 52 (1994), 789.
44. J.G. Zhou and Z. He, *RP. J.*, Vol. 5, (2) (1999), 82.
45. *DTM's Selective Laser Sintering Process and The Sinterstation 2000 System*, DTM Corporation, Austin, TX, 1994.
46. D.L. Bourell, H.L. Marcus, J.W. Barlow and J.J. Beaman, Intern. *J. Powder Metall.*, Vol. 28, (4) (1992), 369.
47. D.L. Bourell, H.L. Marcus, J.W. Barlow, J.J. Beaman and C.R. Deckard, *Multiple Materials Systems for Selective Laser Sintering*, 1990, U.S. Paten #4944817.
48. D.H. Abbott and F.G. Arcella, *Adv. Mater. Proc.*, Vol. 153, (5) (1998), 29.
49. F.G. Arcella, *Titanium' 92 Sci. Tech.*, Ed. F.H. Froes and I.L. Caplan, Warrendale, PA., TMS, 1992, 1395.
50. F.G. Arcella, *41st Sagamore Confer. Proc.*, Ed. W. Roy, Watertown, MA, ARL, 1994, 145.
51. F.G. Arcella and F.H. Froes, *JOM*, Vol. 52, (5) (2000), 28.
52. Z.D. Chen, T.A. Mai and G.C. Lim, *Confer. Proc. ICALEO'98*, Orlando, Florida, USA.
53. P.A. Vetter, T. Engle and J. Fontaine, *Laser Mater. Proc.: Industrail and Microelectronics Applications*, Bellingham: Spie–Intern. Soc. Optical Eng., 1994, 452.
54. J. Lin and W.M. Steen, *J. Laser Applications*, Vol. 10 (1998), 55.
55. J.Y. Jeng, S.C. Peng and C.J. Chou, *Intern. J. Adv. Manuf. Tech.*, Vol. 16 (2000), 681.
56. N.K. Tolochko, T. Laoui, Y.V. Khlopkov, S.E. Mozzharov, V.I. Titov and M.B. Ignatiev, *RP J.*, Vol. 6 (2000), 155.
57. D.H. Abbott and F.G. Arcella, *Adv. Mater. Proc.*, Vol. 5 (1998), 29.
58. G. K. Lewis and E. Schlienger, *Mater. Design*, Vol. 21 (2000), 417.
59. J. Mazumder, J. Choi, K. Nagarathnam, J. Koch and D. Hetzner, *JOM*, May (1997), 55.
60 M. Shellabear, " Benchmark Study of Accuracy and Surface Quality in RP Models ", *RAPTEC, Task 4.2 Report* 2, June 1999.
61 P.E. Reeves and R.C. Cobb, *Proc. of 5th Eur. Confer. on RP. and Manuf.*, Helsinki, (1996), 59.
62 F. Xu, F, *Ph.D. Thesis*, National University of Singapore, 2000.
63. T.T. Wohlers, *Proc. of The 8th Intern. Confer. on RP, Tokyo*, (2001), 22.

Chapter 2

FUNDAMENTALS OF LASER-LITHOGRAPHY PROCESSES

2.1 Laser-lithography

Laser-Lithography (LL), also known as Stereo-Lithography (SL), is one of the earliest and most widely used RP techniques. This process is a combination of computer graphics, laser technology and photochemistry. Using this method, a designed part is synthesized via photo-polymerization of a photosensitive monomer, such as acrylate or epoxy. Since its early development, it was recognized that LL could be applied to many aspects of the manufacturing process. The scope of application of prototype parts generated by LL has grown quite quickly beyond its initial application of rapid 3-D concept model fabrication for visualization and design verification. Despite the great advantage of this technology, the LL process is still not well understood nor controlled. Shrinkage, distortion and relatively weak mechanical properties, which affect the accuracy and quality of the LL part, remain major problems hindering the wider application of this technology. Understanding the fundamentals of the LL process is therefore important to improve the process and expand its applications.

2.2 Laser scanning in LL process

In the LL process, an ultra-violet (UV) laser beam is employed to selectively harden successive thin layers of photo-curable resin such as acrylic-based polymer, building each on top of the previous layer until a three-dimensional part is formed (Fig. 2.1). Each layer in a built-up part is fabricated via a six-step process [1]; leveling, deep dip, elevation, sweep, Z-wait and laser scanning. Once the laser scanning is completed, the above six steps are repeated for the next layer until the whole 3-D part is fabricated. Upon completion of part fabrication, the elevator is raised above the surface of the resin in the vat and excess resin from within the structure of the part is allowed to drain back into the vat. The drained part is then removed manually from the elevator. The laser scanned "green" part is then rinsed using a suitable solvent (e.g. Tri-Propylene Glycol Monoethyl Ether (TPM)) to remove any uncured resin from its surface.

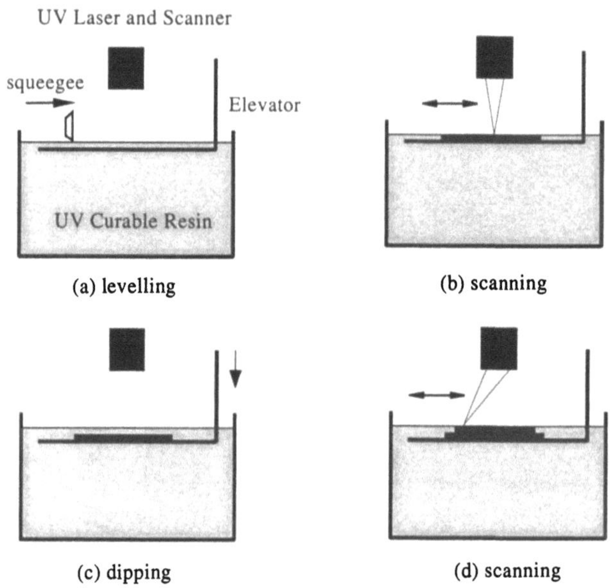

Fig. 2.1 A typical fabrication process using LL method.

2.2.1 Laser-lithography systems

Fig. 2.2 shows a typical LL system that includes a laser system (UV or He-Cd laser), a X-Y scanner (Galvano mirrors), acoustic optical modulator (AOM), Z-focus lens, workstation or PC, machine controller, squeegee, etc. The JSC-2000 [2] uses a high-power Argon laser with a maximum power of 500mW. The laser is powered by an AC 200V (3-phase) 45A supply and the temperature of the laser tube is regulated by cooling water with a flow rate of 8.5 liters/min.

Acoustic Optical Modulator (AOM)

The AOM is a high-speed modulator that is used to rapidly switch the laser beam on or off. It operates on the principle that the refractive index of the optical material is changed by acoustic waves, causing the laser to be deflected away from the vat when not needed.

Z-focus Lens

The z-focus lens moves back and forth to maintain the focus on the resin surface as the laser scans across the stage.

X-Y Scanner

This is a high-precision deflector device that is comprised of two small low-inertia galvanometers mounted with mirrors. The mirrors are arranged 90° to one another so that each is responsible for scanning along one direction. The laser is reflected from the mirrors onto the stage. The coordinated movement of both mirrors directs the focal spot to trace out the desired raster or vector patterns.

Squeegee

This is the recoater blade that traverses across the vat to sweep excess resin from the uppermost layer of the part. The uppermost surface of the previously cured layer is positioned one layer thickness below the bottom edge of the recoater blade. Once the blade completes its motion, the system is ready for the next step.

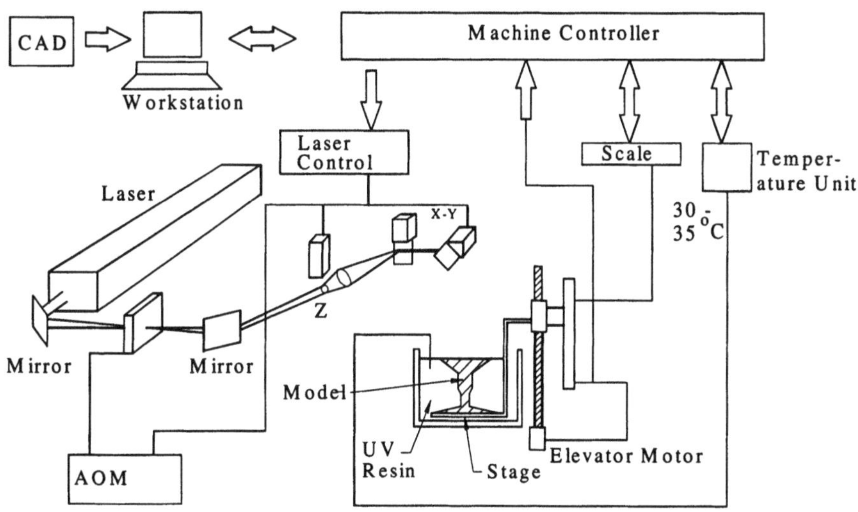

Fig. 2.2 Hardware Setup of JSC-2000 [2].

2.2.2 Laser systems

One of the most important properties of the laser used for the LL process is its transverse mode, which determines the intensity distribution across the width of the laser beam. The simplest or "lowest order" transverse mode has a smooth beam profile with a peak in the middle. Its shape is that of the Gaussian curve.

The first mode is called the TEM_{00} mode (Transverse, Electric and Magnetic modes). Other types of mode TEM_{mn} can be obtained by controlling the optical losses within the laser cavity between the laser mirrors. These losses can be introduced by using an optical limiting aperture within the laser tube. By decreasing the size of the aperture, the TEM_{00} mode will dominate. The TEM_{00} mode is the most desirable for LL because the spread it experiences from diffraction approaches a theoretical minimum value (Gaussian beam).

Argon Laser

Light is emitted when atoms fall to lower energy states after they have been ionized through electric discharges and set into highly excited states, Ionization of Argon requires high density excitation energy. Therefore, a large current is applied onto a very thin tube to increase the current density. A high magnetic field is also applied to prevent diffusion of electrons to the wall of the thin tube so as to increase the power of the laser (Fig. 2.3). However, the efficiency of an Argon laser is very low and the ion temperature of the plasma in the tube goes up to 3000°K. Hence, UV Argon laser units are large and require water-cooling. The output end of the tube is sealed with the Brewster window that is made of crystalline quartz. This is to minimize the loss of energy in the tube, and provide a vertically polarized output beam.

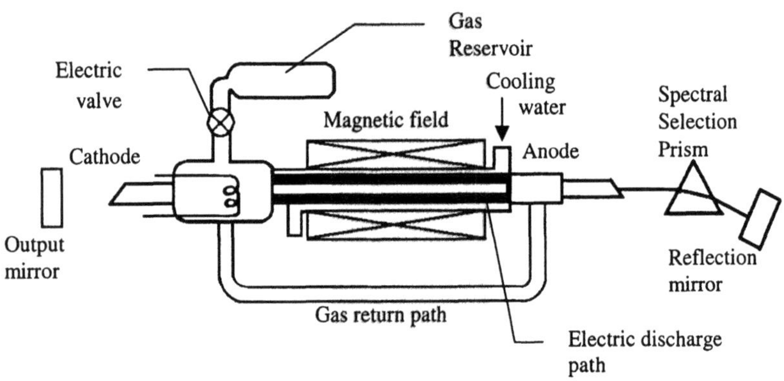

Fig. 2.3 Argon laser configuration (TEM_{00} mode: power = 0.4W and wavelength = 333.6 - 363.8 nm) [3].

He-Cd Laser

Helium (He) atoms are promoted to a semi-stable level with electric discharge. When they collide with Cadmium (Cd) atoms, the He atoms are ionized and the Cd ions at high energy level are excited to emit light of wavelength 441.6 nm (visible) or 325 nm (UV). The He-Cd laser (Fig. 2.4) produces lower energy density, lower discharge current, and lower temperature compared with Argon laser. As Cd is solid at room temperature, the Cd reservoir has built-in heater to vaporize Cd (250~300°C). Since Cd becomes liquid before vaporization, the laser tube cannot be tilted or used upside down. Start-up time takes 10 minutes to reach stable vapor state of Cd. The power of the laser is lower (around 100 mW) than that of the Argon laser.

Semi-conductor Excited Solid-state Laser

The transformed wavelength of the YAG laser is excited by a high power semi-conductor laser using KTP (KT_iOPO_4) crystal. Halving the wavelength of this SHG (Second Harmonic Generation) green laser using non-linear optical crystal produces UV wavelength of 266.1nm. The alignment of the optical components is very critical and it is difficult for the output to remain stable. With recent improvements, this laser has become stable and is now being used as the light source for some RP systems.

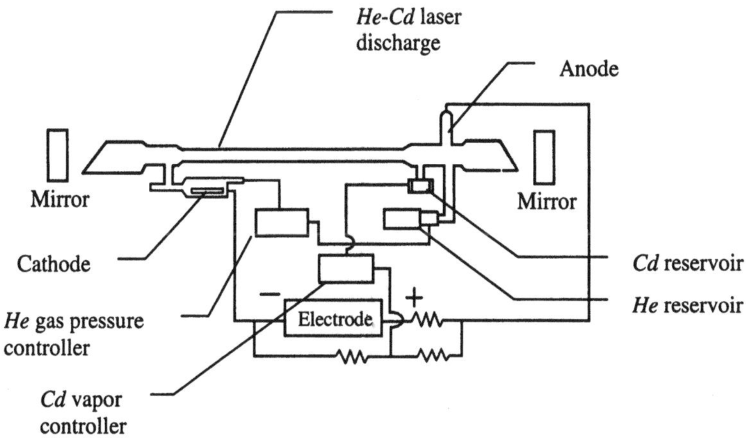

Fig. 2.4 He-Cd laser configuration [3].

2.2.3 Control software

Most of LL systems include three types of software: (i) interface software, (ii) data processing software, and (iii) model creation software. The JSC-2000 interface software supports the IGES, STL, and HPGL formats. Design models can be created by any CAD systems and then translated to the STL format. The data processing software is used to set the optimal orientation for part building and to slice the CAD model to the required layer thickness. The software allows the user to review the sliced data and to modify any open paths detected in the sliced data and the addition of supporting structures to be built along with the model. The model creation software is used to control the part-building process. Operating parameters must be appropriately defined within the program in order to create a part accurately. Model scaling and multiple model creation are also supported by the program. Some of the important process parameters included in the LL software will be described later.

2.3 Fundamental relationships

2.3.1 Working curve

The laser exposure $E(z)$ within the resin varies according to the Beer-Lambert exponential law of absorption [1]:

$$E(z) = E_0 \exp\left(-\frac{z}{D_p}\right) \tag{2.1}$$

where E_0 is the laser exposure on the surface of the resin (mJ/cm^2); D_p is the "penetration depth" (μm) of the resin or the depth where the laser exposure is 1/e of that at the surface; and z is the depth (μm) from the surface of the resin.

For photopolymers, when the exposure is less than a critical value E_c, the resin remains liquid (see Fig. 2.5). When the exposure is greater than E_c, the resin polymerizes. When the exposure is equal to E_c, the polymer is at "gel point", which is the transition from liquid phase to the solid phase. Considering that the laser beam has a Gaussian distribution in the surface coordinate orthogonal to the scan direction and the maximum centerline laser exposure incident to the surface is E_{max}, then the maximum cure depth of a single laser cured line is given by:

$$C_d = D_p \ln\left(\frac{E_{max}}{E_c}\right) \tag{2.2}$$

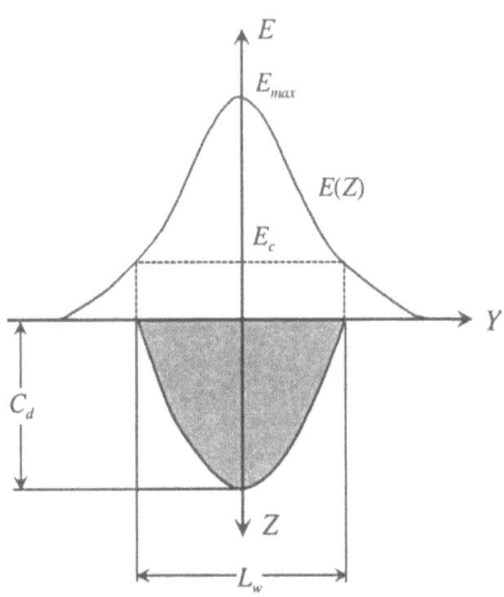

Fig 2.5 Energy distribution on laser exposure and cured resin.

A plot of C_d versus ln E_{max} should give a straight line with a slope equal to D_p and an exposure intercept equals to E_c. This equation is known as the "working curve equation". Since D_p and E_c are purely resin parameters, both the slope and the intercept of the working curve are therefore independent of the laser power.

If the laser beam is a Gaussian beam scanning at a constant speed of V_s over the surface of the resin, the laser exposure function can be shown as [1]:

$$E(y,z) = \sqrt{\frac{2}{\pi}}\left(\frac{P_L}{W_0 V_s}\right)\exp\left(-\frac{2y^2}{W_0 V_s}\right)\exp\left(-\frac{z}{D_p}\right) \tag{2.3}$$

where y is in the distance from the center line (μm); P_L is the laser power (mW); W_0 is the $1/e^2$ Gaussian half-width (μm); and V_s is the laser scanning velocity (mm/s). The maximum exposure should occur along the centerline of the scan (y=0) at the resin surface (z = 0). Thus it is given by substituting y = 0 and z = 0 in Eq. (2.3). Also,

$$E_{max} = \sqrt{\frac{2}{\pi}}\left(\frac{P_L}{W_0 V_s}\right) \tag{2.4}$$

It follows from Eq. (2.3) that for $E(y,z) = E_c$, y is maximum when $z = 0$. Substituting Eq.(2.4) into Eq.(2.3), and letting $z = 0$ and $2y_{max} = L_w$, we obtain:

$$L_w = \sqrt{2}\, W_0 \sqrt{\ln\left(\frac{E_{max}}{E_c}\right)} \tag{2.5}$$

This equation is also known as the "cured line-width function" [1].

2.3.2 Intensity profile

Fig. 2.6 shows the 3-D and 2-D laser intensity profiles. Consider a laser beam being scanned in a straight line at a constant velocity over the liquid polymer. The curing profile is analyzed with the following three assumptions [1]: (i) the photo-polymer resin obeys the Beer-Lambert's law of exponential absorption; (ii) the laser irradiance distribution is Gaussian; and (iii) the resin transition from the liquid phase to the solid phase is at the "gel point".

$2W_0$ is the beam diameter for a Gaussian beam, defined as the beam diameter at H_0/e^2. The coordinate system is first defined. The laser is scanning along the x-direction. The y co-ordinate is in the lateral scanning direction and is orthogonal to the laser scan axis, with the positive direction defined by the right hand rule. Thus y is equal to zero directly under the centerline of the laser scan axis. The z-axis extends directly downwards from the laser scan axis and is measured normal to the x-y plane of the free resin surface where z = 0.

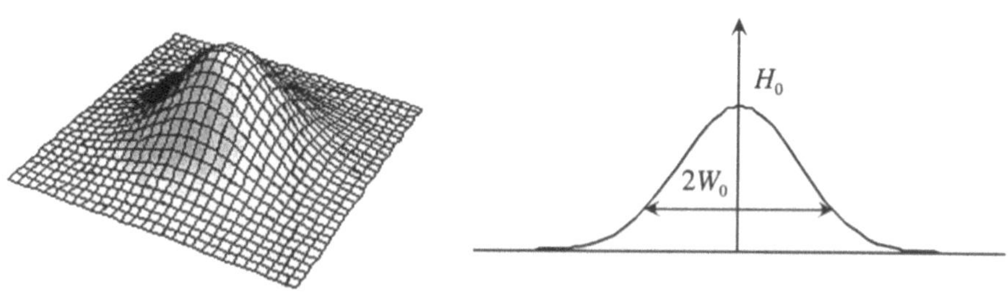

Fig. 2.6 The Gaussian intensity profile.

From Beer-Lambert's law,

$$H(x,y,z) = H(x,y,0)\,e^{-\frac{z}{D_p}} \tag{2.6}$$

where $H\,(x,\,y,\,z)$ is the irradiance at an arbitrary point, $H\,(x,\,y,\,0)$ is the surface irradiance, and D_p is the penetration depth. Further, since the laser irradiance distribution is assumed to be Gaussian, i.e.

$$H(x,y,0) = H(r,0) = H_0\,e^{-\frac{2r^2}{W_0^2}} \tag{2.7}$$

where W_0 is radius of the Gaussian beam ($1\,/\,e^2$) and H_0 is the peak radiance. To define H_0, the total laser power incident on the resin surface (P_L) must be equal to the integral of the laser irradiance distribution over the entire resin surface, i.e.

$$P_L = H_0 \int_0^{\infty} H(r,0)2\pi r dr = 2\pi H_0 \int_0^{\infty} e^{-\frac{2r^2}{W_0^2}} r dr \tag{2.8}$$

Solving from the peak irradiance at the free resin surface, H_0, the following result is obtained [6].

$$H_0 = \frac{2P_L}{\pi W_0^2} \tag{2.9}$$

Substituting this value into Eq. (2.7), and then Eq. (2.6), the Gaussian laser irradiance distribution function in cylindrical coordinates is obtained:

$$E(y,\,z) = \int_{-\infty}^{\infty} H(\,r(\,t\,),\,z)\,dt \tag{2.10}$$

$$V = \frac{dx}{dt} \tag{2.11}$$

$$r^2 = x^2 + y^2 \tag{2.12}$$

The exposure $E(x,\,y,\,z)$ by a line scan of laser beam can be calculated from:

$$\begin{aligned} E(y,z) &= \frac{2P_L}{\pi W_0^2 V_s} e^{-\frac{z}{D_p}} e^{-\frac{2}{W_0^2}y^2} \int_{-\infty}^{\infty} e^{-\frac{2}{W_0^2}x^2} dx \\ &= \sqrt{\frac{2}{\pi}} \frac{P_L}{W_0 V_s} e^{\left(-\frac{2y^2}{W_0^2} - \frac{z}{D_p}\right)} \end{aligned} \tag{2.13}$$

By integrating the laser beam irradiance moving at a constant velocity V along the x-axis, the maximum centerline laser exposure on the surface of photo-polymer, E_{max}, can be calculated by substituting $y = 0$ and $z = 0$ into Eq. 2.3.

$$E_{max} = E(0,0) = \sqrt{\frac{2}{\pi}} \frac{P_L}{W_0 V_s} \tag{2.14}$$

Here the critical exposure value, E_c is introduced. If the exposure is less than E_c, the polymer remains liquid; and if greater than E_c, the polymer solidifies. Also,

$$\frac{2}{W_0^{\ 2}} y^2 + \frac{z}{D_p} = \ln\left(\frac{E_{max}}{E_c}\right) \tag{2.15}$$

where $E(y,z) = E_c$ represents the border of a cured line profile. Since W_0, V_s, E_c are all constants, the equation shows that the cross-section of a single cured line has a parabolic shape, as shown in Fig. 2.7. Defining the curing width, C_w (or line width L_w) and the curing depth, C_d, of the parabolic profile of one scanned line of photo-polymer, the curing depth is the value of z when $y = 0$ and half of the curing width is at the value of y when $z = 0$. Therefore, C_w and C_d are represented by the working curve equation shown in Eq.(2.2).

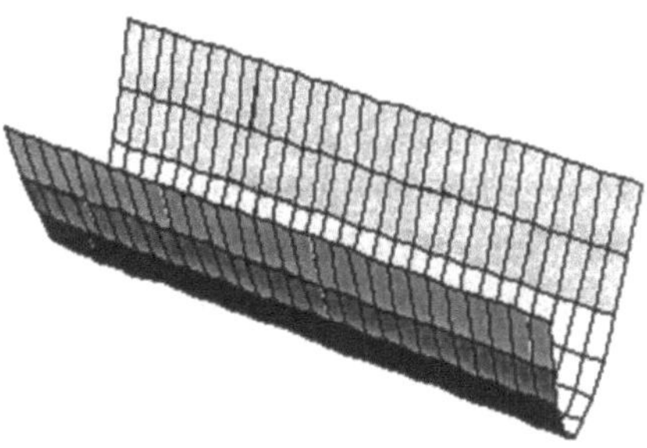

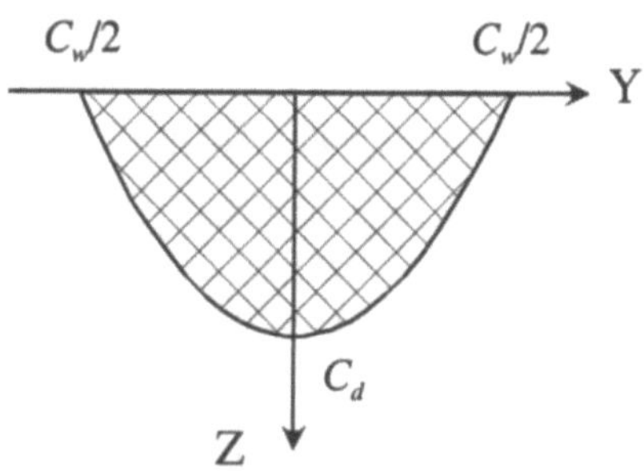

Fig. 2.7 Single cure line profile.

2.3.3 Process parameters

It has been reported that there are as many as 30 parameters affecting the LL process [4]. However, a recent study conducted by Schaub *et al.* [5] reported that more than 50 process parameters are involved in a single resin and part family type. Some of the main parameters affecting the accuracy of the built part are shown in Table 2.1. Among them, the laser power, layer pitch and scan pitch parameters influence the mechanical properties of the part built significantly. This will be examined and discussed further in the next chapter.

Table 2.1 List of important LL process parameters

Parameters	Function
Step size	Controls the speed of laser scanning.
Scan delay	Specifies the time the scanner waits before drawing the next vector
Jump size	Controls the speed of the laser movement when not drawing
Jump delay	Specifies the settling time for the scanner after executing a jump
Laser on	Synchronizes the laser to the phase delay of the scanner for precise commencement of scanning
Laser off	Synchronizes the laser to the phase delay of the scanner for precise line ending
Laser power	Determines the energy exposure onto the material
Step period	The time required for the scanner to move the specified step size
Scan offset	Compensates for the laser beam size at the boundaries of the scanned sections
Scan pitch	Controls the distance between adjacent lines of the scan
Layer pitch	Specifies the layer incremental distance during part building

Some of the main parameters affecting the accuracy of the built part that are worthy of attention are as follows [7]:

(a) Accuracy and Integrity of STL CAD Data

When converting the CAD model to STL format (the standard file format for RP machines), surfaces are discretized into facets (tessellation). The chosen size of the facet determines the accuracy of the built part using the tessellated data [8, 9].

(b) Flatness of the Building Platform

As the parts are built layer-by-layer on the stage, any deviation of the stage surface flatness will induce inaccuracy in the part. Furthermore, if the stage surface is not leveled, part accuracy would also be affected.

(c) Initial Position of Stage

The laser beam is calibrated at a fixed distance (900mm) from the free resin surface to produce an accurate *X*-*Y* scan. However, if the initial position of the stage is above the free resin surface or the resin level falls below the calibration level, it would result in error in the *Z* direction (vertical) of the model.

(d) Laser Characteristics

The laser beam diameter may change slightly as the laser tube ages. There may also be shift in the optical alignment due to small temperature variations and the mode structure of the laser may be altered. This would in turn alter the cured polymer characteristics.

(e) Changes in Photopolymer Properties

In the normal course of part building, the photopolymer parameters or characteristics may vary upon refilling of the vat (tank in which part is built) with photopolymer from a different lot. Also, photopolymer properties may vary with aging.

(f) Variation of Laser Beam Size across Surface of Vat

The shape and diameter of the laser beam spot change from the center of the vat to the outer edge. The laser beam spot changes from a perfect circular spot at the center to an elliptical spot at the extreme position. In addition, the beam will also move out of focus as the angle increases. Unless these changes are dynamically compensated by software, the resulting errors may be significant (error as large as 1.770 mm has been reported [10]).

(g) Changes in Photopolymer Level during Building

The resin surface level must be maintained at the correct *Z*-level for optimal laser focus. However, during polymerization, the photopolymer undergoes about 5% to 7% volumetric shrinkage. Of this amount, 50% to 70% occurs in the vat [1]. Because of this, the photopolymer surface level will deviate from the optimal level during fabrication, thereby affecting the laser focus.

(h) Settling Time (Z-wait) for Photopolymer before Scanning

Once the platform has moved to the required position, in principle, the photopolymer on the top of the previous layer should blend seamlessly with the free surface after the action of the squeegee (see Fig. 2.1a). Unfortunately, due to surface tension effect, a meniscus will be formed on top of the layer, above the free surface. If sufficient time is allowed for the resin to undergo fluid dynamic relaxation, this effect will decay [1]. Otherwise, the layer built will be thicker than desired. The relaxation time is dependent upon the photopolymer viscosity and surface tension.

(i) Accuracy of the Squeegee

For accurate building, the lower edge of the squeegee should be leveled with the free surface of the resin (Fig. 2.1a).

(j) Laser Control Parameters

Most RP model creation software requires users to specify the various laser control parameters e.g. Laser ON/OFF, delay, scan offset, etc. If these are not properly set, accuracy will be affected.

(k) Dimension Changes due to Shrinkage

The shrinkage upon polymerization results in deviation of dimensions of the part from the nominal values.

(l) Distortion during Building

The non-uniform shrinkage of each layer and the complex inter-laminar shear stresses result in distortion of the part in the form of warpage and curl. The distortion may be significantly reduced if the part is properly supported.

(m) Post-cure Distortion

It has been reported that post-cure distortion is related to the percentage of uncured photopolymer left within the part after scanning [11-13]. This is because the part would be subjected to further shrinkage and the resulting inter-laminar effect would lead to more distortion during post-curing.

(n) Delamination

Delamination is the separation of the layers or detachment of the profile scanned line from the raster-filled area. Layer separation is due to the incorrect setting of the layer pitch for a given laser exposure. It may be also caused by the inter-layer distortion. Profile line detachment is due to improper setting of the "Laser On", "Laser Off" and "Scan Offset" parameters.

(o) Entrapped Bubbles

Entrapped bubbles are a common source of error during part building. During prolonged building, tiny bubbles diffuse from the photopolymer. These bubbles would be coated on the top of the previous layer of the part when the squeegee sweeps across the vat and be trapped in the part after scanning by the laser. When the squeegee motion is too fast, bubbles will also be formed as air will also be trapped during separation of the resin from the edge of the squeegee.

(p) Trapped Volume Effect

Trapped volume occurs in parts with enclosed geometry such that pockets of liquid resin within the interior of the part are isolated from the rest of the liquid in the vat. As the trapped liquid resin cannot be drained to the vat, it forms a steady meniscus above the previous scanned layer. The meniscus is not readily eliminated by the squeegee action and therefore causes the built layer to be thicker than required. If this effect is not rectified, it may result in the squeegee hitting the built part.

(q) Swelling Effects

A part tends to swell and lose dimensional stability on prolonged exposure in the vat (about 20 hours) caused by a change in the cross-link density. This is due to a chemically induced solubility change where the system absorbs itself, causing differential shrinkage between areas with different laser exposure. Because of the change in the cross-link density, swelling is also accompanied by the softening of the part and a corresponding reduction in strength. The degree of swelling is dependent on the degree of curing of the part [1].

(r) Distortion after Support Removal

Distortion also occurs after improper removal of the support structures. During part removal, the applied force may induce distortion to the part. The removal of the support structures may also induce relaxation on the "stress-frozen" part and re-distribute the stress. This will inevitably be manifested as distortion.

2.3.4 Profile of scan lines

2.3.4.1 Single-scan line

The study of the characteristics of a single-line cured resin is crucial to a better understanding of the photo fabrication process. This is because the single line of cured resin is the most basic element in part building, as each layer in a built-up prototype is made up of a successive series of single-line cures.

Fig. 2.8 shows the cross-section of a single line of cured LL resin that can be formed by sweeping the laser beam across the surface of a vat of liquid resin. In the previous figure (Fig. 2.7), a schematic to represent the cross-sectional

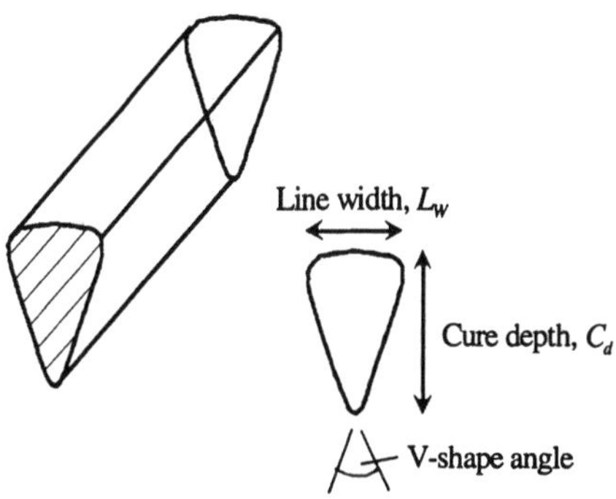

Fig. 2.8 Cross-section of single cure line.

profile of a single line of cured resin together with the intensity profile of a "Gaussian" laser beam is also shown. As can be observed from the schematic, the intensity of the laser beam decays radially from the center of the laser spot. Due to this and the fact that no polymerization occurs at laser intensities lower than the critical intensity, E_c, the depth to which polymerization takes place varies radially from the center of the laser spot. Judging from the schematic, one would expect the depth of polymerization to be greatest at the center of the laser spot and will not occur at any point where the intensity is less than E_c. At radial points where the laser intensities are equal to E_c, the polymer is at the "gel point" corresponding to the transition from the liquid phase to the solid phase. Hence, the general shape of the cross-section of the single line of cured resin follows closely to the shape of the intensity profile of the laser beam as shown early.

Processes

Fig. 2.9 illustrates the experimental set-up used to determine the characteristics of a single-line of cured resin. A piece of glass slide is placed on the building platform and lowered until its top surface is flushed with the free surface of the resin. The laser is set to scan multiple lines across the glass slide in such a way that a portion of the scanned lines protruded from an edge of the glass slide. The slide is then removed from the platform and the pieces of cured lines were dried using a piece of soft paper towel. A sharp razor blade is used to trim the protruding ends of the cured lines. Trimming is necessary as the scanned lines grow thicker and wider toward their ends, due to the deceleration of the laser beam at the end of each scan. The cross-section of each scanned line is then observed under a microscope fitted with a camera that is linked to a thermal printer. The cure depth C_d and the line width L_w of each cured line are then measured from a hardcopy produced using the thermal printer by comparing it to a printout of a graduation scale taken under the same magnification. The entire procedure is repeated for lines generated under different laser power settings and various scanning speeds.

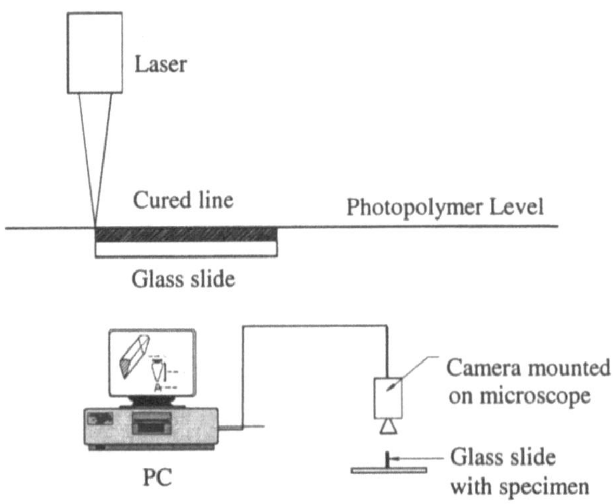

Fig. 2.9 Experimental set-up for determining the profile of a single-line scan.

For the experiments carried out to evaluate the SCR300 resin [14], 24 sets of data were collected for different laser power settings and scanning speeds. For each set of data, about 10 specimens of the cured "lines" were collected. The illumination power, which is the actual power of the laser exposing on the resin surface, was measured using a power meter for each laser power setting. This was done so that any loss in laser intensity across the optical system was accounted for when calculating laser exposure.

Fig. 2.10 and 2.11 show the relationships between C_d and E_{max} and between L_w and E_{max}, using the relationships given in Eq. (2.2) and (2.5), respectively. The results obtained adhere closely to the Beer-Lambert law of exponential absorption and the laser beam can be modeled as having a Gaussian intensity distribution profile. This is due to the fact that only a Gaussian laser beam scanned at a constant velocity across a resin obeying the Beer-Lambert Law will result in a single cured line with a parabolic cross-section. Using least square fitting, the following empirical relationships were derived for the SCR300 resin:

$$C_d = 195.22 \ln (E_{max}) - 121.96 \qquad (2.16)$$
$$L_w^2 = 64028 \ln (E_{max}) - 77{,}516 \qquad (2.17)$$

Since the intercept of the ln (E_{max}) axis of Eq. (2.2) is equal to E_c,

$$195.22 \ln (E_c) = 121.96$$
$$E_c = 1.86 \text{ mJ/cm}^2$$

Also from Eq. (2.16), $D_p = 195.22\mu\text{m}$.

Hence, the laser intensity required to cause polymerization in the SCR300 resin was found to be 1.86 mJ/cm^2. The gradient of Eq. (2.2) gives the penetration depth D_p , which is constant for a particular type of resin. Eq. (2.2) and (2.5) can be used to predict both the cure depth C_d and the line width L_w of a built layer respectively, if the laser exposure density at the surface of the resin E_{max} is known.

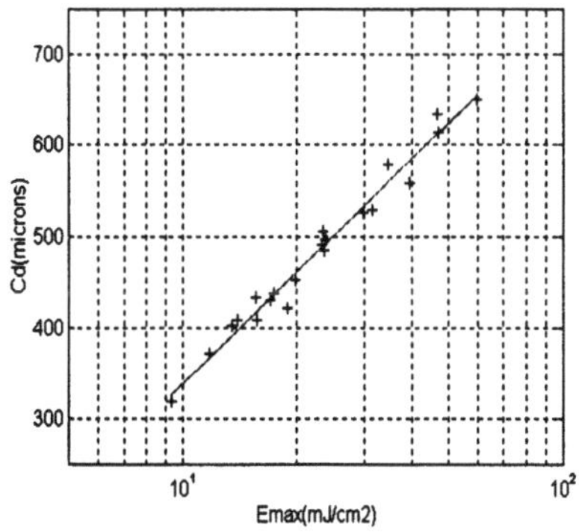

Fig. 2.10 Working curve for SCR300 resin.

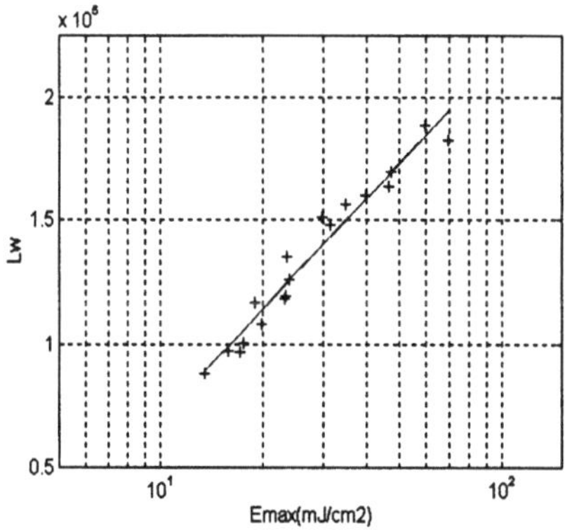

Fig. 2.11 Cured line width with laser exposure for SCR300 resin.

The cross-sectional profile of a single-layered part with multiple scanned lines is a function of laser power, scanning speed, and the other factors which govern the properties of laser-cured parts [10]. The single line specimen is adopted because it represents the fundamental element of the photo-fabricated component. Fig. 2.12 shows the typical cross section of a cured single-layered part. Since the laser energy is partially absorbed and partially reflected by the resin during laser scanning, the upper portion of the cured line absorbs much more energy than that at lower portion such that the cross section of the line appears in a V-shape. It is known that the laser exposure is a function of the laser power and the scanning speed used. To characterize the influence of both laser power and laser scanning speed on the shape of a cured part, Eq. (2.4) can be used. As shown in Fig. 2.13, the depth and width of a cured line, which are mainly the function of laser exposure for a fixed scanning pattern, increase with the increase in laser exposure.

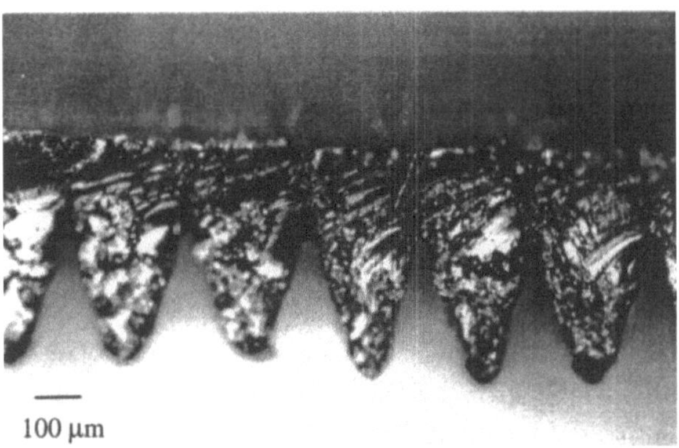

Fig. 2.12 Cross-section of single-cured layer with overlapping scan pitch of 200μm.

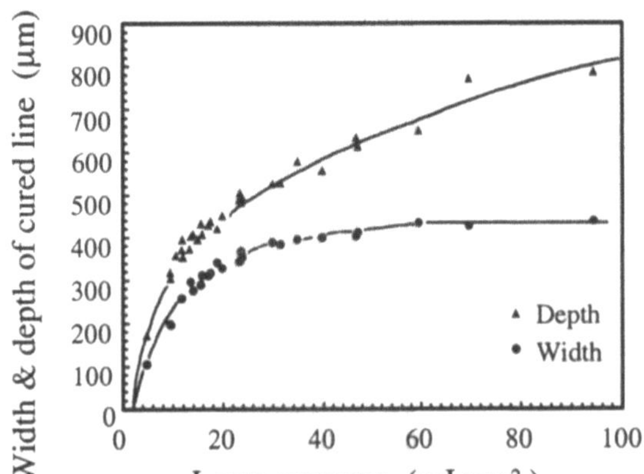

Fig. 2.13 Cured line width and depth vs. laser exposure.

Besides laser exposure, the laser spot size also contributes to the profile of a cured line. The slope of the curve reaches the maximum value when the exposure is about 4 mJ.cm^{-2} indicating that the polymerization rate is maximum and decays with the exposure energy as shown in the figure. Physically, it is understood that when the laser beam penetrates into the liquid resin, its energy is partially absorbed and partially reflected by the resin resulting in a decrease in its energy. When the penetrating laser energy is less than the minimum value required for polymerization, the curing process stops. Hence, the depth of a cured line is laser energy dependent. However, the width of a cured line is determined by the laser spot size.

The relationship between the width and depth of a laser-cured line is shown in Fig. 2.14. It is noted that at the lower laser exposure range, an increase in the width of a cured line is a linear function of the depth. However, the width will approach a constant value corresponding to the spot size of the laser beam if the depth of the cured line is further increased. As a result, the V-shape angle of a cured line is consequently decreased. This relationship is graphically shown in Fig. 2.15. Initially, the V-shape angle increases with the laser exposure. However, the V-shape angle reaches its saturation value when the intensity of laser exposure is about 60 mJ cm^{-2} where the depth of a cured line still increases with the laser exposure while the width of the cured line reaches its saturation value. Therefore, an optimal laser exposure can be determined accordingly. It is noted that the V-shape angle is an important parameter in laser curing.

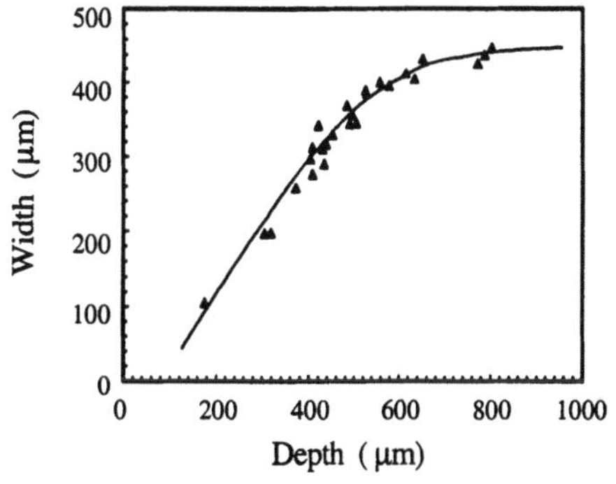

Fig. 2.14 Cured line-width vs. line-depth.

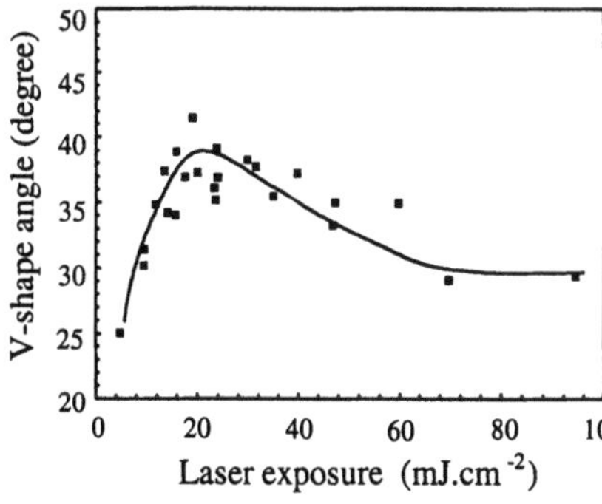

Fig. 2.15 The change of V-shape angle.

2.3.4.2 Multi-scan line and layer

The cross-sectional profile of the multi-layered part is illustrated in Fig. 2.16. Examination of the cross-section of the cured part revealed three areas according to the degree of curing (Fig. 2.17). The areas identified are the *over-cured*, *under-cured* and *uncured* areas. The uncured areas are made up of liquid resin entrapped within the structure of the part during the layer-by-layer building process. The existence of different degrees of cure at different positions in the cross-section of the single line is due to the decay in the intensity of the laser beam when it penetrates into the resin. Therefore, since the upper portion of the single-scan line is exposed to higher laser intensity, it has a higher degree of cure compared to the lower portions. Furthermore, when the Nth layer is built on top of the $(N\text{-}1)^{th}$ layer, the laser beam may penetrate into the upper portion of the $(N\text{-}1)^{th}$ layer leading to further curing (over-cured region). In addition, when a layer is built, an overlap between two adjacent scanning passes is needed to provide a continuous connection across the layer. As a result, the overlapped areas are further cured (in some cases, it is not fully cured depending on the laser power), while the degree of cure in other areas remains unchanged.

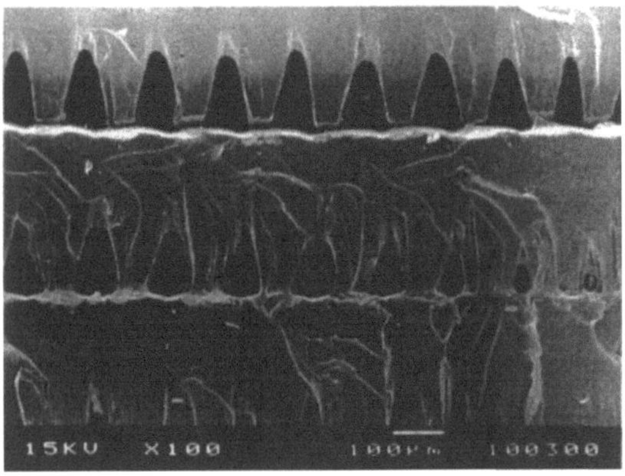

Fig. 2.16 Morphologies of multi-scanned layer with a layer pitch of 300μm.

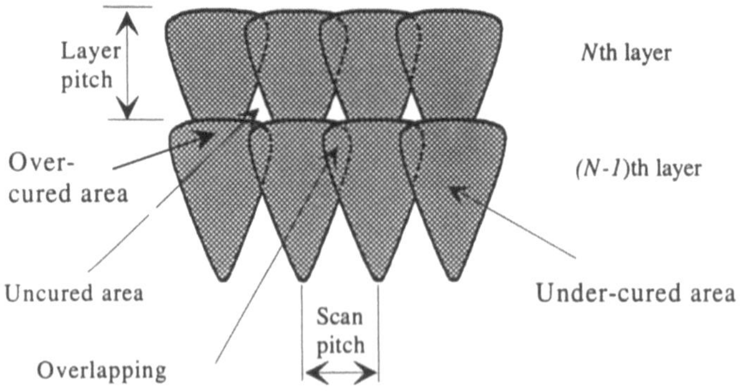

Fig. 2.17 Cross-sectional profile of multi-layer structure [16].

Simulation work was performed at a laser power setting of 250 mW and a step size of 20 μm for the SCR300 resin. Fig. 2.18 shows the result of the simulation compared to photographs of the test specimens taken using a camera attached to an optical microscope. It can be observed from the results that the model under-estimates the cure depth of the specimens by about 100 µm for layers formed by overlapping scan lines. This may be due to the fact

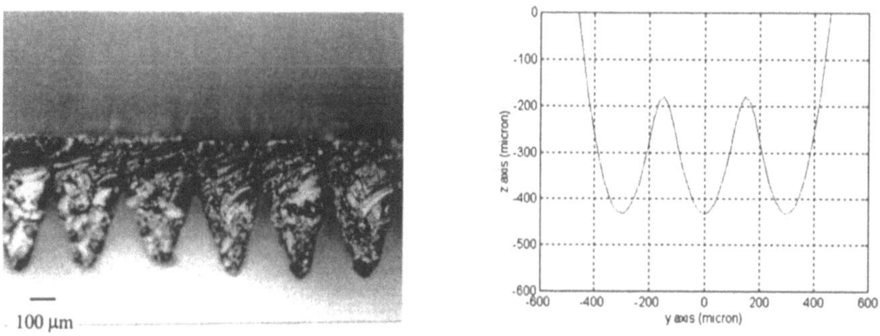

(a) No overlapping, scan pitch = 311µm

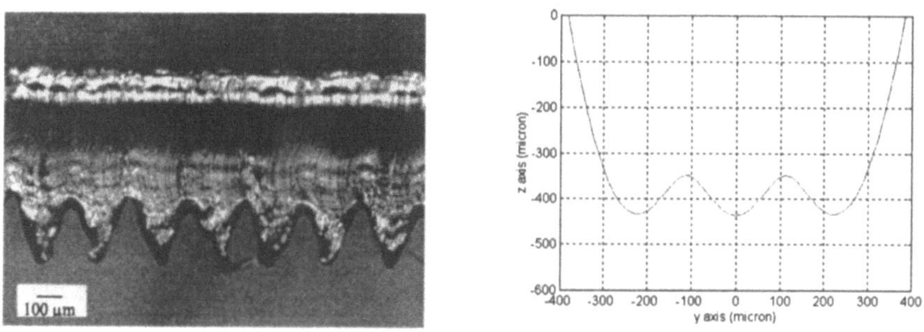

(b) 25% overlap, scan pitch = 233µm

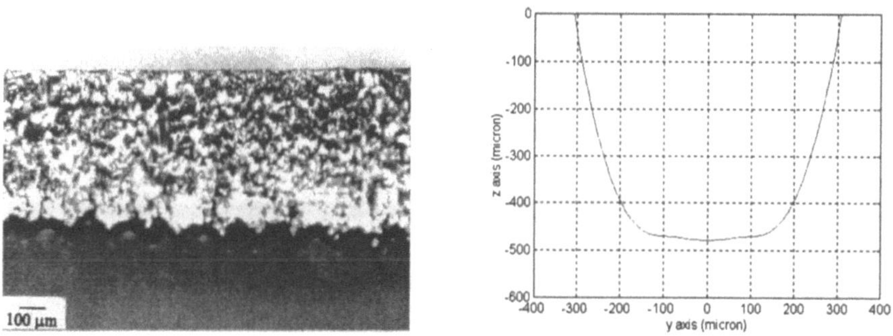

(c) 50% overlap, scan pitch = 156µm

Fig. 2. 18 Comparison of experimental and simulation results for single layer building.

that the simulation model does not consider the additional curing induced by the laser beam when it penetrates the solidified portion of the previously cured line. However, the layer profiles produced by the simulation are similar to the actual profiles of the built prototypes. Hence, the simulation is capable of providing a qualitative indication of the layer formation process. By reducing the scan pitch on both the simulation model and the actual built model, it can be seen that the cured profile becomes more uniform due to the increased amount of overlapping between adjacent scan lines. It was found that at a 50% overlap where the scan pitch is half the value of L_w, the fabrication of a layer with uniform cure depth is possible. The advantage brought about by having a uniform cured profile is evident in the following chapter that deals with the mechanical and physical properties of the LL prototypes.

2.4 Characteristics of photo-polymerization

2.4.1 Properties of photo-polymer

Photo-curable resins are made of three main components: (i) a photo-initiator that generates free radicals upon exposure to ultra-violet (UV) radiation; (ii) a monomer or pre-polymer end-capped with acrylate groups; and (iii) a reactive acrylic diluent to lower the viscosity of the resin. Typical photo-curable resin formulation contains 2% to 5% of photo-initiator and equal parts of acrylic prepolymer and diluent [17].Usually known as photo-polymers, photo-curable resins can be solidified by a UV light source. The resin is composed of photo-initiator (I) and series of monomers (M), oligomer (O), other molecules (T) and some amount of diluent. The reaction scheme is:

Initiation of reaction: $I \xrightarrow{hv} I\bullet \rightarrow I\bullet$ (radical)

Growing reaction: $I\bullet \rightarrow M$ (or O) $\rightarrow I\text{-}M\bullet + M$ (or O) $\rightarrow I\text{-}M_n\bullet$

Chain reaction : $I\text{-}M\bullet + T$ (other molecular) $\rightarrow I\text{-}M + T\bullet$ (radical shifting)

Stop reaction: $I\text{-}M\bullet + T\bullet \rightarrow I\text{-}M\text{-}T$

$I\text{-}M\bullet + I\text{-}O\bullet \rightarrow I\text{-}M\text{-}O\text{-}I$

where h is Planck's constant, v is the frequency of the photon and hv is the photon energy. The photo-initiator is activated by UV energy. The activated radical activates the monomer to start the growing reaction. It then moves to the other molecules to initiate the chain reaction. When two radicals combine together, the reaction is terminated.

Oligomer

The mechanical property of a model depends heavily on the characteristics of the oligomer. Most photopolymer uses urethane or epoxy acrylate as oligomer. The oligomer of SCR-310 is urethane acrylate with molecular weight in the range of several thousands. Generally, epoxy acrylate does not have much shrinkage problem but urethane acrylate does. Recently, the properties of both types of acrylate have been improved with less shrinkage and higher tenacity.

Photo-initiator

Aromatic carbonyl compounds are commonly used as the photo-initiator in radical polymerization. Different photo-initiators have different absorption spectrum of wavelengths. Thus, photo-initiators suitable for the wavelength of the source laser must be selected. The curing depth can be controlled by the transparency of the photo-polymer. A sensitizer can be added to the photo-initiator to increase its sensitivity.

Monomer

The molecular weight of a monomer is around several hundreds, and is much smaller than the oligomer. Therefore, it plays the role as a diluent. The monomer also increases the sensitivity of polymerization and accelerates the cross-linking to prevent swelling of the model.

Diluent

This serves as additional chemical to improve the performance of the photo-polymer, such as lubricant agent, preservative, or dyes.

2.4.2 Photo-polymerization

Polymerization is the process of linking small molecules (monomers) into larger molecules (polymers). Photo-polymerization is one such process that is initiated by exposure to radiation. It consists of three essential steps, namely *initiation, propagation,* and *termination.* Radicals (indicated as •) are groups of atoms that cannot exist independently. However, they may exist in combination with other molecules (in this case, *R*), to form free radicals, which can then exist independently. A radical catalyzes chemical reactions, which affect the rest of the molecule, while maintaining its own identity.

Photo-polymerization is an exothermic process and is thus energy efficient. Its efficiency allows productive use of low-power UV lasers in rapid prototyping. Upon exposure to UV radiation, free radicals ($R\bullet$) are formed from the photo-initiator (I) upon absorption of photons of certain frequency. The free radicals react with a monomer (M) to initiate a chain reaction [1].

$$I + h\nu \rightarrow 2R\bullet \tag{2.18}$$

$$R\bullet + M \rightarrow (R - M)\bullet \tag{2.19}$$

In going from weak Van der Waals forces between monomer molecules to a network of covalent bonds in the polymer, many bulk properties are changed.

As the shear strength increases, the resin changes from liquid to solid. The average distance between groups decreases, resulting in an increase in density (i.e. shrinkage).

2.4.3 Over-curing and over-penetration

Normally, a single scanned line cannot be fully cured and the degree of curing at different positions in the cross-section of the single line is not identical (refer to Chapter 3 for details) because the intensity of laser power diminishes and the laser is partly reflected when it penetrates into the resin. Therefore, the upper portion of the single scanned line has a higher degree of cure while the lower portion is less cured.

There are two reasons for over-curing. The first reason is that the laser power is normally strong enough to cure deeper than the desired cure depth. Thus, when curing the bottom-most layer of an over-hanging part, below which is liquid resin rather than the stage or the previously cured layer, we will get a layer thicker than the desired layer pitch. Fig 2.19 compares the actual cure depth d and the desired layer pitch L_p of the cured layer of a non-overhanging part (left) and an overhanging one (right).

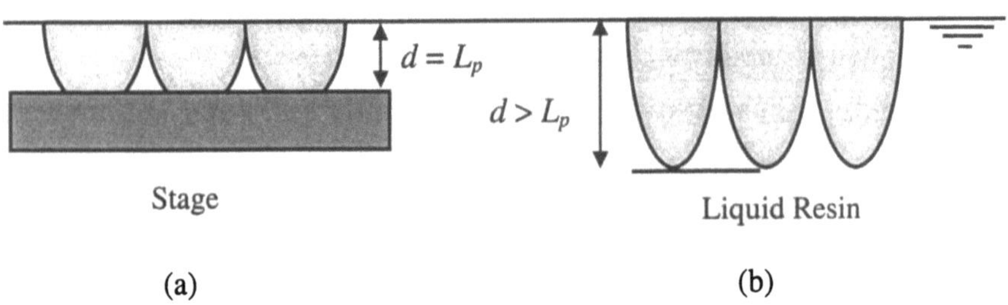

Fig 2.19 Over-curing effect.

Another reason is that when curing the second layer, the liquid resin below the previously cured layer is also cured. This is due to the over-penetration of the laser, suggesting that the solid resin is sufficiently transparent to the UV laser to allow the laser to penetrate through to the lower layer.

Modeling of Over-Penetration

Reference to Fig 2.20, d_1 is the cure depth of the first layer, and d_2 is the total cure depth for the first and second layers. After each scan, the stage will be lowered by the amount of L_p. Suppose that the laser does not over-penetrate the first layer, then in principle:

$$d_2 = d_1 + L_p \tag{2.20}$$

However, in actual case, the laser does over-penetrate the first layer. Thus:

$$d_2 > d_1 + L_p \tag{2.21}$$

In this section, we shall derive the equation for the cured depth of an *n*-layer model.

(1) *First layer*

From Eq.(2.2), the cured depth d_1 of the first layer can be obtained:

$$d_1 = D_p \ln\left(\frac{E_{max}}{E_c}\right) \tag{2.22}$$

At the bottom of the first layer, the laser energy is just enough to cure the resin. From Eq. (2.1), the energy there is given by:

$$E_c = E_{max}\, \mathrm{e}^{-\frac{d_1}{D_p}} \tag{2.23}$$

In addition to D_p which has been defined as the depth of penetration by the laser through the liquid resin, we shall define D_p^s as the depth of penetration through the solid resin.

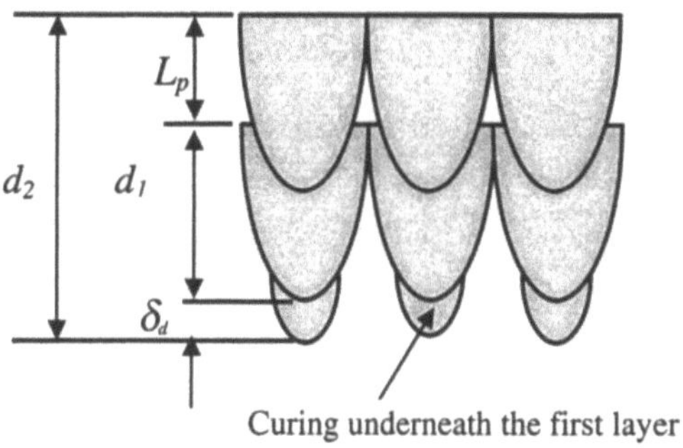

Fig. 2.20 Effect of over-penetration leading to further over-curing [19].

(2) *Second layer*

The laser energy at the bottom of the previously cured layer, which is now L_p + d_1 below the free resin surface, is given by:

$$E(L_p + d_1) = E_{\max}\, \mathrm{e}^{-\frac{L_p}{D_p}} \mathrm{e}^{-\frac{d_1}{D_p}} \tag{2.24}$$

If this energy is larger than E_c, then there is over-penetration. This leads to an over-cured depth of δ_d below the first layer.

At the bottom of the δ_d-thick 'layer', the laser energy is equal to E_c:

$$E(L_p + d_1 + \delta_d) = E_{max}\, e^{-\frac{L_p}{D_p}} e^{-\frac{d_1}{D_p^S}} e^{-\frac{\delta_d}{D_p}} \tag{2.25}$$

Taking logarithm, the above equation can be written:

$$\ln\left(\frac{E_{max}}{E_c}\right) = \frac{1}{D_p}(L_p + \delta_d) + \frac{d_1}{D_p^S} \tag{2.26}$$

Multiplying both sides by D_p and also from Eq. (2.22), d_1 can be obtained:

$$d_1 = D_p \ln\left(\frac{E_{max}}{E_c}\right) = L_p + \delta_d + \frac{D_p}{D_p^S} d_1 \tag{2.27}$$

From Fig. 2.20, it can be seen that:

$$d_2 = L_p + d_1 + \delta_d \tag{2.28}$$

Substituting Eq. (2.27) into Eq. (2.28), it follows:

$$d_2 = d_1 \left\{1 + \left(1 - \frac{D_p}{D_p^s}\right)\right\} \tag{2.29}$$

If the energy given by Eq. (2.24) is smaller than E_c, then:

$$d_2 = L_p + d_1 \tag{2.30}$$

(3) *Third layer*

The laser energy at the bottom of the previously cured layer, which is now L_p + d_2 below the free resin surface, is given by:

$$E(L_p + d_2) = E_{max}\, e^{-\frac{L_p}{D_p}} e^{-\frac{d_2}{D_p}} \tag{2.31}$$

If this energy is larger than E_c, then there is over-penetration. This leads to an over-cure of δ_d below the first layer. At the bottom of the δ_d-thick 'layer', the laser energy is equal to E_c:

$$E(L_p + d_2 + \delta_d) = E_{max}\, e^{-\frac{L_p}{D_p}} e^{-\frac{d_2}{D_p^S}} e^{-\frac{\delta_d}{D_p}} \tag{2.32}$$

Taking logarithm, it follows:

$$\ln\left(\frac{E_{max}}{E_c}\right) = \frac{1}{D_p}(L_p + \delta_d) + \frac{d_2}{D_p^S} \tag{2.33}$$

Multiplying both sides by D_p and substituting Eq. (2.22), it follows:

$$d_1 = D_p \ln\left(\frac{E_{max}}{E_c}\right) = L_p + \delta_d + \frac{D_p}{D_p^S} d_2 \tag{2.34}$$

But, we can see that:

$$d_3 = L_p + d_2 + \delta_d \tag{2.35}$$

Thus, substituting Eq. (2.34) into Eq. (2.35), and then substituting Eq.(2.29) in the resulting equation, it follows:

$$d_3 = d_1 \left\{ 1 + \left(1 - \frac{D_p}{D_p^S}\right) + \left(1 - \frac{D_p}{D_p^S}\right)^2 \right\} \tag{2.36}$$

If the energy given by Eq. (2.31) is smaller than E_c, then:

$$d_3 = L_p + d_2 \tag{2.37}$$

(4) *N-th layer*

In general, if $E(L_p + d_{n\text{-}1}) > E_c$, then there is over-penetration. The thickness of the n-layer model will be given by:

$$d_n = d_1 \sum_{k=1}^{n} \left(1 - \frac{D_p}{D_p^S}\right)^{k-1} = \frac{D_p^S}{D_p} d_1 \left\{ 1 - \left(1 - \frac{D_p}{D_p^S}\right)^n \right\} \tag{2.38}$$

Otherwise, the thickness of the n-layer model will be given by:

$$d_n = L_p + d_{n-1} \tag{2.39}$$

If Eq. (2.38) is useful until the m-th layer, then Eq. (2.39) is modified to:

$$d_n = (n - m)\, L_p + d_m \tag{2.40}$$

Fig. 2.21 compares the actual and ideal thickness of models with different number of layers. The ideal thickness of an n-layer model is simply L_p multiplied by n. The curve portion of the actual thickness curve corresponds to Eq. (2.38) while the straight portion corresponds to Eq. (2.39) and should have the same gradient as the ideal thickness line. This means that after a few layers, the error will become constant. Table 2.2 shows the experimental data which tallies well with the modeling and its prediction.

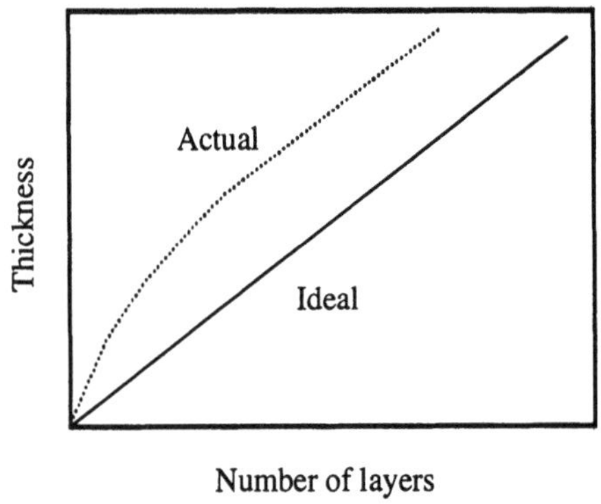

Fig. 2.21 Actual and ideal thickness versus number of layers.

Table 2.2 Experimental data for thickness versus number of layers

No. of layers / (resin type)	Thickness (mm)	
	SCR-300	SCR-310
1	0.415	0.390
2	0.795	0.690
3	1.075	0.965
4	1.290	1.195
5	1.495	1.410
6	1.700	1.635
7	1.955	1.855

2.4.4 Focus effect on the curing profile

The width of the laser beam is known to be hyperbolic, with the spot size being minimum at the focal point, also called the beam waist (Fig. 2.22). Curing is done at the defocused area and the approximation gives a reasonable result. The focal point for normal curing is usually 8mm below the surface of the photo-polymer, and the spot diameter is 240 μm.

To obtain the intensity of a single scan line of the laser beam, integrate $H(t,y,z)$ with t from $-\infty$ to ∞.

$$E(y,z) = \int_{-\infty}^{\infty} H(vt, y, z)dt \tag{2.41}$$

The photo-polymer is solidified where $E(y,z)$ exceeds E_c, the critical energy to solidify the photo-polymer. Thus the area (y,z) which satisfies:

$$E(y,z) > E_C \tag{2.42}$$

is the curing profile. Based on Eq. (2.41), the energy exposure distribution is simulated using Mathematica [18], as shown in Fig. 2.23 and 2.24.

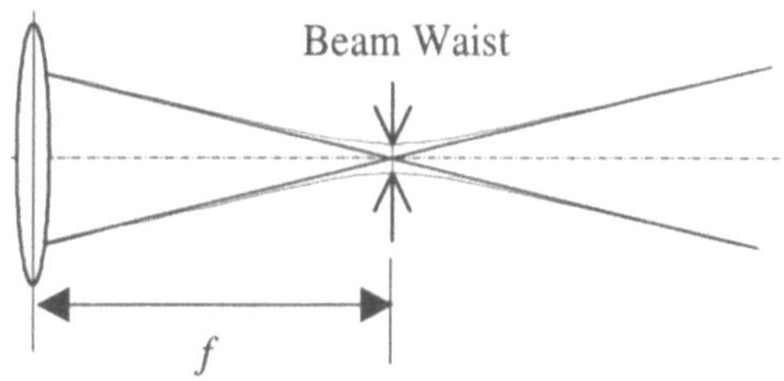

Fig. 2.22 Focusing and beam waist.

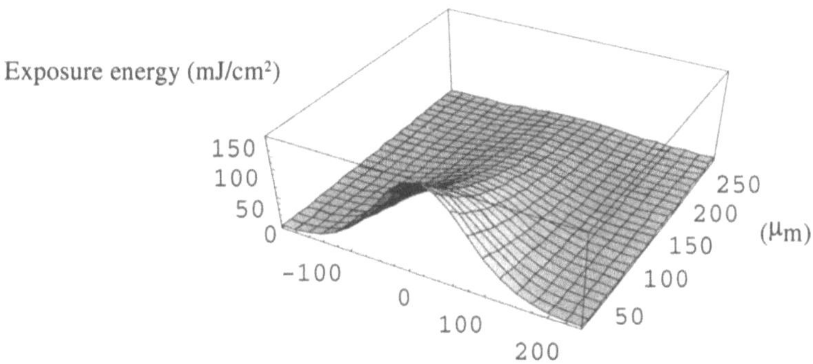

Fig. 2.23 Energy exposure for a single line scanning.

If the focal point is near to the photo-polymer surface, the curing profile becomes a triangular shape (Fig. 2.24d). If the focal point is deep enough, the shape of the curing profile is parabolic and there is no focal effect except the changing of beam diameter at the surface of the photo-polymer. i.e. if $z = 0$, the curing width is only effected by the beam diameter and not by the focal effect.

Experiments on the curing depth with different focal depths have been reported [6] (see Table 2.3). The laser beam was scanned in one line with: laser power on the surface of 17 mW; and scanning speed of 747 mm/sec. Each scan was done at different positions of the platform. The curing profile was observed in the same way as discussed in section 2.2.1. The result shows that simulation considering the focal beam is close to the actual curing profile (Fig. 2.25). As mentioned before, however, the focal point is usually not very near to the photo-polymer surface, i.e. around 8 to 10 mm. In this condition, not much of the focal effect needs to be considered.

Table 2.3 Experimental data on curing depth with different focal depths.

Focal Depth (mm)	Curing Depth (μm)
10	260
9	280
8	295
7	340
6	375
5	410
4	495
3	550
2	740
1.5	800

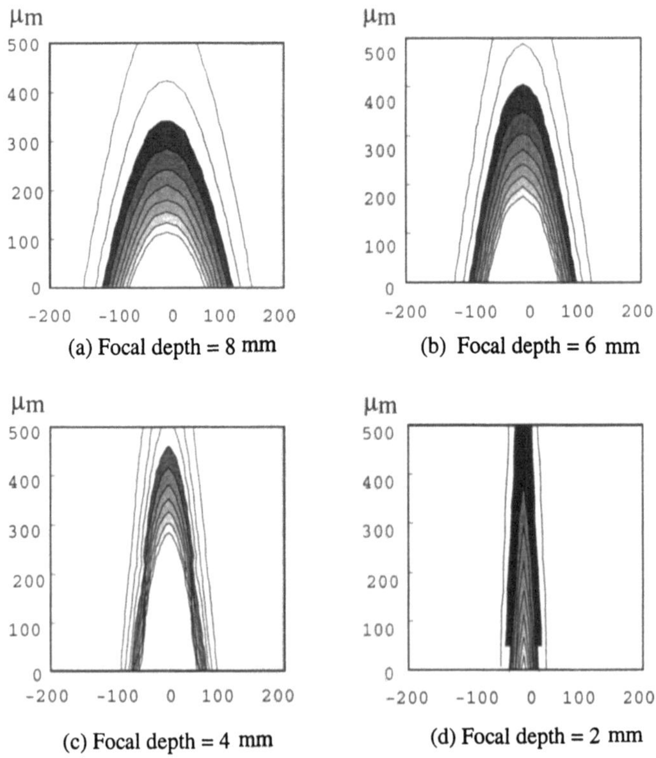

Fig.2.24 Simulation of focus effect on curing profile [6].

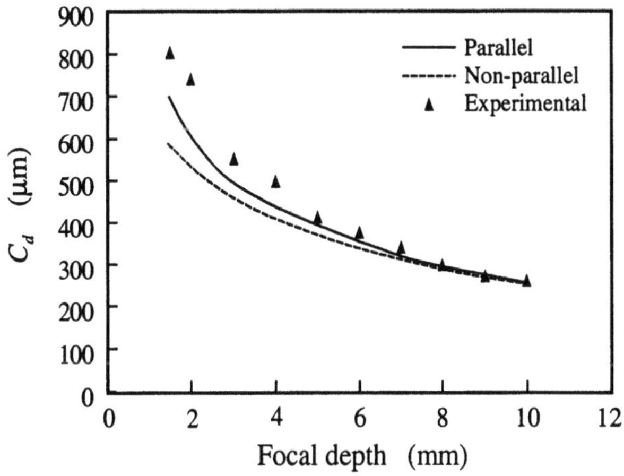

Fig. 2.25 Simulated and experimental curing depth.

2.5 Summary

This chapter introduces the basic theory and modeling of the laser-lithography (LL) process. The fundamental behaviors of photo-polymerization are methodically discussed. Since the single scan line and cured layer are the basic building blocks for any LL parts, an understanding of the chemistry and mechanics will help to explain the causes of shrinkage and distortion of LL-built parts and will lead to better control of the LL process. Due to the uncontrollable over-curing and over-penetration effects during the laser scanning process, the LL process has been relying very much on optimized process parameter setting. The modeling presented in the chapter explains some of the part quality issues. The detailed experimental study to verify the findings is discussed in the next chapter.

2.6 References

1. P.F Jacobs, *RP. & Manuf. : Fundamentals of Stereolithography*, Society of Manufacturing Engineers (SME), Dearborn, MI, 1992.
2. K. Arita, A. Ikeda, and J. Kuzusako, *Proc. of 1991 Intern. Confer. on Industrial Electronics Control & Instrumentation*, IECON '91/IEEE.
3. Laser Association, *Laser Application Technology Handbook*, Asakura Press, Tokyo, Japan (1984), 18. (in Japanese).
4. J.P. Kruth, *Annals of CIRP*, Vol. 40 (1991), 603.
5. D.A. Schaub, K.R. Chu, D.C. Montgomery, *J. Manuf. Syst.*, Vol. 16, No. 4 (1997), 290.
6. T. Miyazawa, *PhD dissertation*, National University of Singapore, 2001.
7. C.M. Cheah, *PhD dissertation*, National University of Singapore, 2001.
8. R.J. Donahue, R.S. Turner, *Proc. of 2nd Intern. Conf. on RP*, Dayton, Ohio (1992), 221.
9. X. Sheng, B.E. Hirsch, *Comp. Aided Design*, Vol 24, No. 8 (1992), 437.
10 O. Bjorke, *CIRP Annuals*, Vol. 40/1 (1991), 1.

11. L Lu, J.Y.H. Fuh, *et al.*, *Mater. Res. Bull.*, Vol.30 (12) (1995), 1561.
12. C.M. Cheah, J.Y.H. Fuh, et al., *J. Mater. Proc. Techn.*, Vol.67 (1997), 46.
13. J.Y.H. Fuh, Y.S. Choo, et al., *J. Mater. Proc. Techn.,* Vol. 63 (1997), 887.
14. Nippon, *SCR-300 Resin Catalogue*, UV curable resin form Nippon Synthetic Rubber Ltd, Japan, 1995
15. L. Lu, J.Y.H. Fuh, et al., *Mater. and Manuf. Proc.*, Vol. 10 (4) (1995), 655.
16. W.L. Wang, C.M. Cheah et al. , *Mater. and Design*, Vol. 17, No.4 (1996), 205.
17. Hoyle, C.E., J.F. Kinstle, *Radiation Curing of Polymeric Materials*, American Chemical Society, Washington, DC., 1990.
18. R. Maeder, *Programming in Mathematica, 3rd Edition*, Addison Wesley Longman, Inc., 1997.
19. J. Toh, *BEng Thesis*, National University of Singapore, 1995.

Chapter 3

MATERIAL CHARACTERIZATION OF LASER-LITHOGRAPHY BUILT PARTS

3.1 Mechanical properties of LL parts

During the last few years, there has been overwhelming global research and development effort towards improving the mechanical properties of LL-built parts. Studies on photo-polymerization [1,2-5] for LL applications aimed at increasing both the curing speed and degree of polymerization. Recent advances in UV-curing chemistry [6-8] focus on developing a low shrinkage resin that shows high efficiency in accelerating the polymerization process so as to improve the mechanical properties of the final product.

Typical stress-strain curves of LL parts are given in Fig. 3.1. It can be seen that high yield stress and large deformation can be obtained at the same time if high laser power and/or small layer pitch are employed. Through several reported experimental studies [9-11], it was found that the mechanical properties of a LL-built part are strongly influenced by the key process parameters set in the LL machine, i.e. laser power and layer pitch.

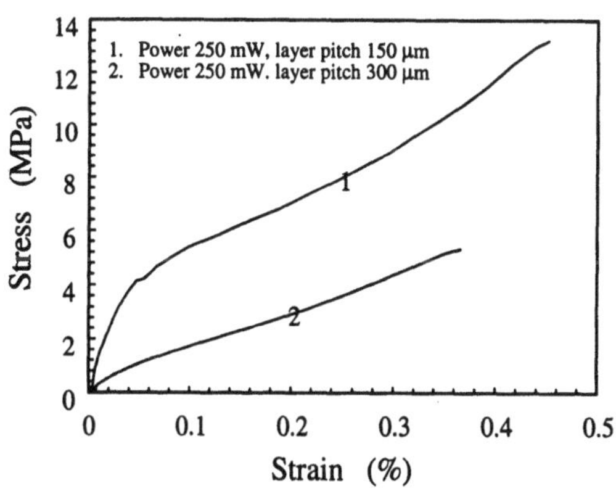

Fig. 3.1 Typical stress vs. strain curves at different operating parameters.

3.1.1 Properties at green state

Rapid prototypes fabricated using the LL process are not fully cured (green-state) after laser scanning. As photo-polymerization is a function of the chemical parameters associated with the photo-initiators, resin components and additives [1], and by technical parameters, such as UV power and wavelength [12], and temperature and environmental conditions, it is very difficult to obtain a fully-cured green state part in practice. In addition, gaps are formed in the prototype during photo-fabrication due to the Gaussian intensity profile of the laser beam, as described in the last chapter. The size of these gaps is dependent on the laser exposure density and the layer pitch used in the fabrication process, as will be explained in the later part of this chapter. These gaps entrap liquid resin from the surroundings during the layer-by-layer building process. The overall structure of the green-state part is hence a combination of cured, partially cured and uncured resin. As a result, a certain degree of in-homogeneity exists within the structure, which depends on the parameters used in the building process.

Fig. 3.2 (a) to (c) show the variation of the modulus of elasticity, ultimate tensile strength (UTS) and the elongation to fracture with laser power for the green-state specimens that were built using a LL machine (SONY Solid Creation System, JSC-2000) [13]. The layer thickness of each specimen was controlled by pre-setting the layer pitch that determines the depth of each successive lowering of the stage on which the specimens were built. The cross-sectional profile of the specimen was created by the laser beam using a single-direction scanning pattern, in which every layer is scanned in the same direction. The laser beam was set to sweep across the surface of the resin at a rate of 1.475 m/s (i.e. scanning speed V_s). The same scanning speed was used to generate all the specimens. All specimens built with dimensions of 50 mm x 10 mm x 4 mm were oriented such that the wiping direction of the squeegee is parallel to the length of the specimens or the tensile direction. The SCR300 photopolymer specimens were built under varying laser powers of 100mW, 150mW, 200mW and 250mW. The variations in the layer pitches were: 150μ m, 200μm, 250μm, and 300μm.

From Fig. 3.2 (a), it can be observed that an increase in laser exposure and a decrease in layer pitch leads to an increase in the elastic modulus of the specimen. This is due to the fact that specimens created using higher laser exposure density and lower layer pitch have a higher degree of curing with less entrapped liquid resin. Since liquid resin does not contribute towards stiffness, only better cured structures will exhibit higher stiffness hence showing an increase in the modulus of elasticity. This factor is critical for part building, as the green-state part should be stiff enough to withstand its own weight and manual handling during and after photo-fabrication.

Fig. 3.2 (b) and 3.2 (c) show the variation of the UTS and elongation to fracture of the green-state specimens with laser exposure. An increase in the laser exposure increases the amount of cross linking between the monomers, resulting in improved strength and toughness of the part, as reflected in the graphs. Similarly, it can be observed that an improvement in the green strength can also be achieved by reducing the layer pitch. This is because the lower layer pitch will result in over-curing of the previously scanned layers causing a higher degree of polymerization in these layers. From the results obtained, it is obvious that a highly cured green-state part has better mechanical strength and can be built by increasing the laser exposure and decreasing the layer pitch. Besides being more durable, having a highly cured green-state prototype will reduce the amount of shrinkage encountered by the part during post-curing and thus preserving its overall accuracy. However, in practice, the layer pitch is often set as large as possible in order to speed up the photo-fabrication process.

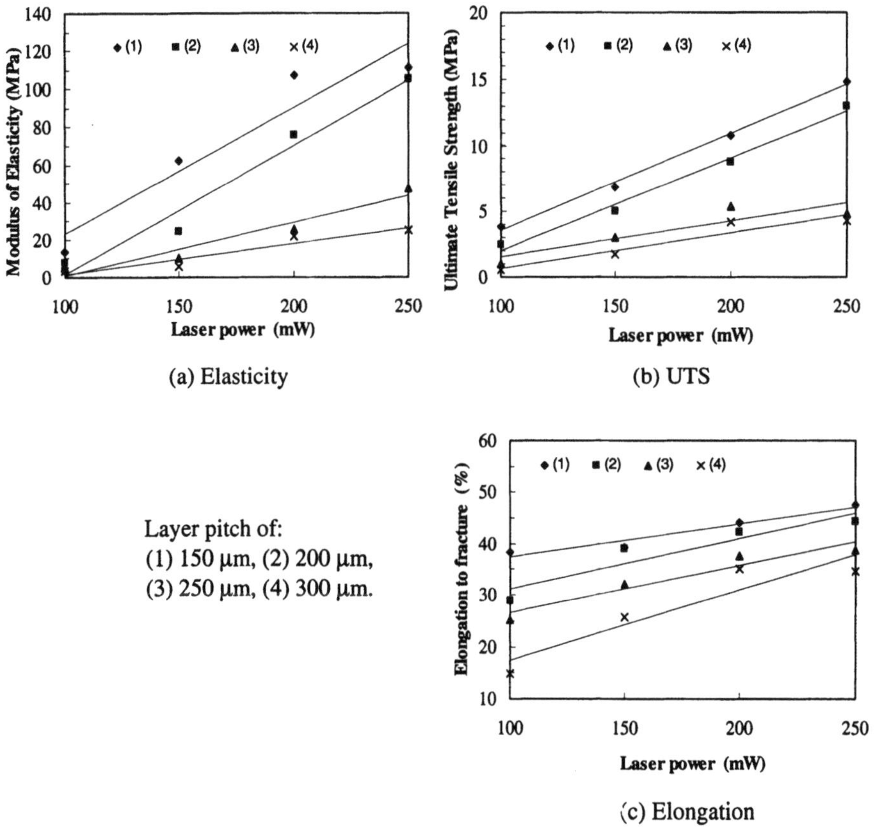

Fig. 3.2 Variation of mechanical properties with laser power and layer pitch, at green state.

3.1.2 Properties after post-curing

In practice, the green parts have to be sent for post-processing that involves part removal, cleaning and post-curing. Post-processing is one of the largest sources of errors in the LL process as it involves changes in the properties of the structure and manual handling [14]. As such, attention has to be given to the steps involved in order to yield an accurately finished prototype.

Although a sufficiently hard green-state part can be built using the correct set of parameters, the prototype has only been partially polymerized giving it a "tacky" surface. It is therefore necessary to post-cure the green-state prototype by exposing it to high intensity ultraviolet (UV) light in a specially built curing unit. Post-curing under high intensity UV light causes the polymerization of the uncured resin retained in the prototype resulting in a fully-cured end product. The improvement in the degree of curing will lead to changes in the mechanical properties as the post-cured part shows improved strength and toughness. This is favorable for the concept of a functional prototype. However, as shrinkage [15,17] and heat release are unavoidable during the polymerization of the retained liquid resin, the model is subjected to distortion [18] and also internal stresses and strains. Hence, it is essential to understand how the mechanical properties of the post-cured prototype are affected by the parameters involved in the fabrication of its green-state counterpart with the aim of building a strong and accurate prototype.

Fig. 3.3(a) to (c) show, at post-cure state, the variations of the modulus of elasticity, ultimate tensile strength (UTS) and elongation to fracture with laser power and layer pitch. Compared to the mechanical properties of its green-state counterparts, an increase in the modulus of elasticity and UTS can be observed. However, it was also noted that the specimens after post-curing encounter a decrease in the elongation to fracture. Fig. 2.17 in the last chapter shows that there is a certain amount of uncured and partially-cured resin trapped within the structure of the green-state specimens that depends on the laser power and layer pitch used. The increase in the elastic modulus and UTS can be understood by the contribution of further polymerization of the uncured and partially laser-cured portions after post-curing. It was noted that although all the specimens were fully cured after post-curing, variations in the mechanical properties with laser power and layer pitch were still prominent.

As the liquid and partially-cured resin polymerizes during post-curing, heat is released due to the exothermic nature of the polymerization. Since photo-curable resins have low thermal conductivity, localized heating may occur leading to the formation of thermal stresses and strains that weaken the structure of the part and also causes distortions in its final shape. Internal stresses can also be formed due to shrinkage of the trapped liquid resin upon polymerization. Shrinkage results in the formation of cavities that lead to discontinuity in the structure of the part. This is evident from the decrease in the elongation to fracture of the post-cured specimens as compared to its green-state counterparts. Since specimens generated with high laser power and low layer pitch contain less entrapped liquid resin, it is expected to show better mechanical properties, as confirmed by the results illustrated in Fig. 3.3.

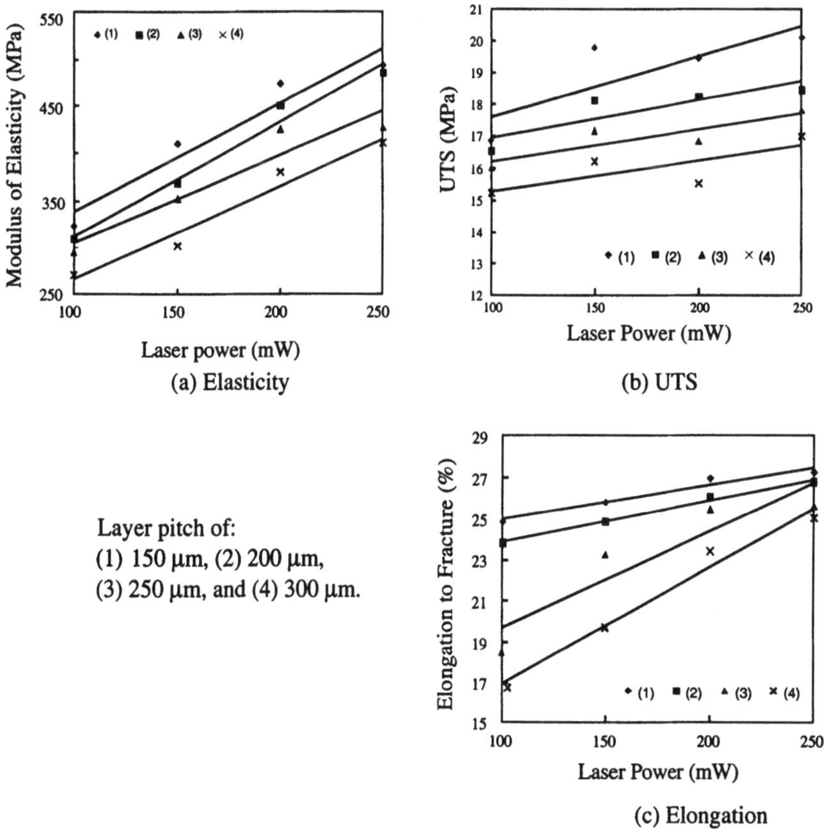

Fig. 3.3 Variation of mechanical properties with laser power and layer pitch, at post-cure state.

3.1.3 Density change

Fig.3.4 shows the density change with different layer pitch and laser power. In general, lower laser power and larger layer pitch result in lower-density parts. Clearly the layer pitch strongly influences the change in density if lower laser power is used. The density cannot be altered very much if a high laser power is used.

3.1.4 Shrinkage and distortion

Photopolymers typically shrink up to 8% by volume during polymerization in the LL process [15]. A detailed study on the shrinkage and distortion can be found in Reference [16]. Shrinkage causes part distortion and warpage due to internal stresses built up during manufacturing, and thus greatly influences the dimensional accuracy. The amount of shrinkage increases with the amount of partially-cured or uncured resin, which in turn depends on the laser power and the layer pitch. Since UV-curable resins with lower shrinkage and improved properties have recently become available, the following therefore address more on the fundamental study of the relationships between the input parameter settings on the LL machine and the resulting directional shrinkages.

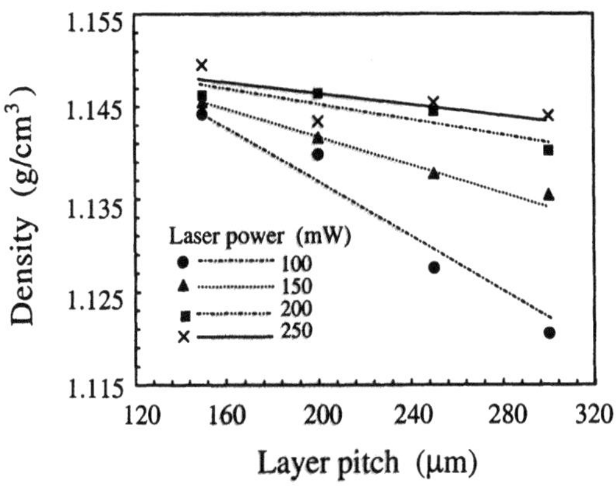

Fig. 3.4 Change in density of laser-cured specimens.

The shrinkage strain is calculated according to the following:

$$\text{Shrinkage}\ (\%) = [\ (D_l - D_p) / D_l\ ] \times 100\% \qquad (3.1)$$

where D_l is the distance between two marks in the scanning or transverse direction before post-curing and D_p, the distance between the same marks after post-curing.

Fig. 3.5 shows the test specimens built by the SONY JSC-2000 LL machine using an acrylic-based photo-curable resin, De Solite SCR-310 [19]. Four different sizes of one layer pitch (i.e. the thickness of one single layer) of 150, 200, 250 and 300 μm were used together with four different laser power settings of 100, 150, 200 and 250 mW, respectively. Four marks were made on the surfaces of each specimen to measure the shrinkage of the specimens in both the scanning direction and transverse directions (normal to the scanning direction) after thermal and UV post-curing. Separate thermal oven and UV chamber were built for the post-curing purpose. After marking, all specimens were post-cured under an UV or heat source.

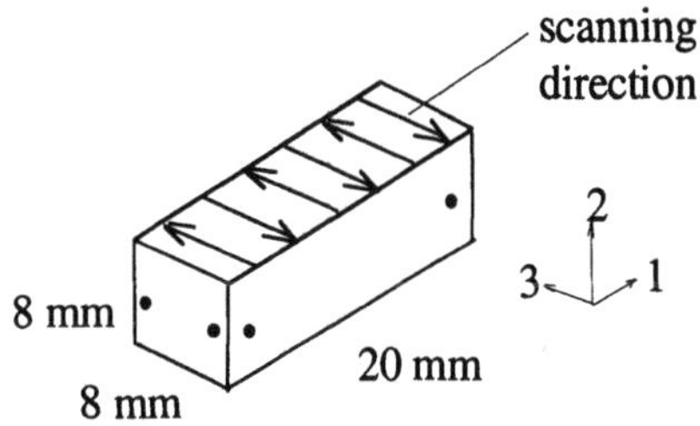

Fig. 3.5 Shrinkage test model.

3.1.4.1 Thermal post-curing

Fig. 3.6 shows an on-line shrinkage measurement in which the specimen has been thermally post-cured. The deformation during thermal treatment consists of thermal dilatation and shrinkage. Because of the difference in the thermal coefficient of the specimen at specific degrees of curing, dilatation due to thermal expansion is different at particular temperatures. At any moment of thermal treatment, the difference Δd between a tangent line and the d versus T curve at any temperature interval ΔT represents the shrinkage due to phase transformation of the specimen. The total shrinkage during heating and cooling is therefore indicated by d_{max} in the figure.

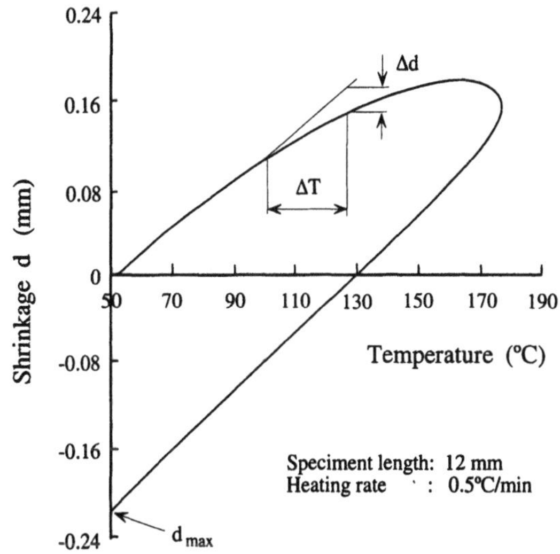

Fig. 3.6 Shrinkage measurement during thermal post-curing [20].

Fig. 3.7 (a) to (d) show the shrinkage trends in the scanning (longitudinal direction-3) and transverse (direction-1 or -2) directions after thermal post-curing. It can be seen from the figures that shrinkages in both directions increase with an increase in layer pitch and with a decrease in laser power. It is noted that shrinkages in the scanning and transverse directions are different. In the case of a small layer pitch and low laser power used in the part building, the shrinkage in the transverse direction is lower than that in the scanning direction. With an increase in layer pitch, the amount of shrinkage in the transverse direction increases faster than that in the scanning direction. Finally, the amount of shrinkage in the transverse direction is larger than that in the scanning direction.

As shown in Fig. 3.7 another observation is that the incremental shrinkage in the transverse direction decreases when a higher laser power is used. When a laser power of 250 mW is used, regardless of the layer pitch, the shrinkage in the transverse direction is always lower than that in the scanning direction. With high laser power, there is reduced amount of un-cured resin in the transverse direction so that the under-cured resin in the scanning direction dominates the directional shrinkage more.

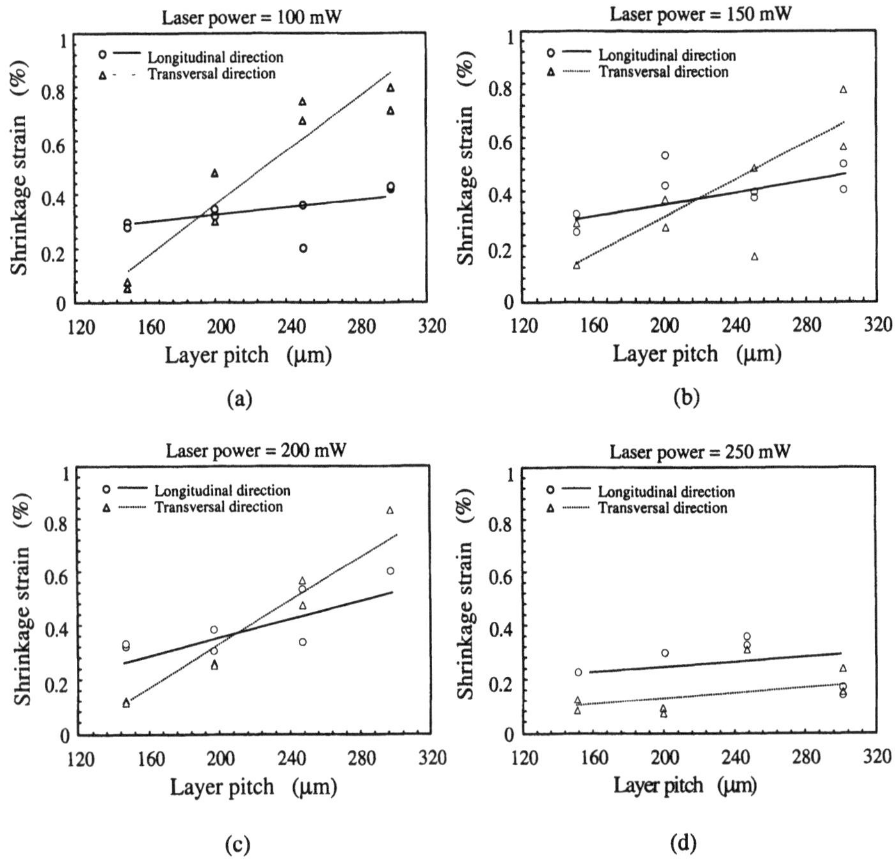

Fig. 3.7 Measure of shrinkage after thermal post-curing, built by different laser powers.

Fig. 3.8(a) and (b) show the in-process measurement of post-cure shrinkage (shrinkage vs. curing time) along the scanning direction at different layer pitches. The curves verify the principle that a lower layer pitch and/or higher laser power will leave less uncured resin, thereby producing less post-cure shrinkage. Firstly, the shrinkage rate is smaller (the gradient is gentler) for the lower layer pitch. Secondly, a higher laser power tends to obscure the effect of layer pitches, since the amount of uncured resin is probably very small and not significantly different for different layer pitches as compared to a lower lascr power.

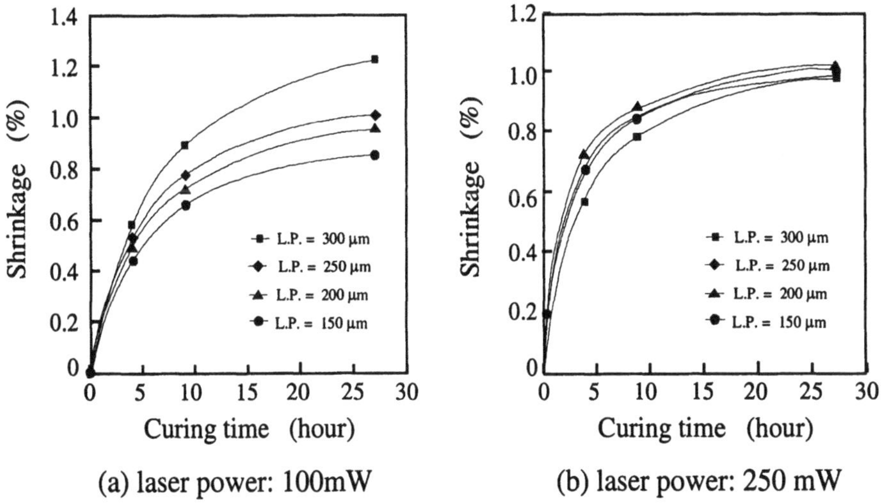

Fig. 3.8 In-process shrinkage measurement from thermal post-curing at 120 degree C [20].

3.1.4.2 UV post-curing

A dynamic measurement technique using embedded sensors was employed by Nau *et al.* [17] who observed that the specimens expanded as they were exposed under UV light but immediately shrank when the UV light was removed. Actually, two reactions take place during post-curing; one of them is thermal dilatation and the other one is shrinkage due to polymerization. The amount of dilatation and shrinkage depends on the degree of curing of a specimen.

For the case of UV curing, the trend is similar to the thermal curing except that the curing rate is much smaller (the gradient is gentler). Fig. 3.9 (a) and (b) show the shrinkage measured after UV post-curing. It also confirms that at the higher laser power and lower layer pitch, the shrinkage will be smaller due to the reduced uncured resin trapped within the multi-layer structure. The shrinkage in the transverse direction is smaller at the lower laser power and lower layer pitch as compared to the shrinkage in the scanning direction, but is always larger at the higher laser power and layer pitch. This is because there is always more under-cured resin trapped in the scanning direction, and this dominates the shrinkage behavior.

It is interesting to note the difference in the shrinkage characteristics from thermal and UV post-curing. As shown in Fig. 3.10, thermal curing, in general, has a higher shrinkage rate than UV post-curing. This is caused by a certain chemical reaction that occurs during the thermal treatment, e.g. loss of vaporized monomer and weight. UV curing, however, only creates cross links among the monomers. Although the curing efficiency is much higher using the thermal treatment, the mechanical properties (elasticity, elongation to fracture, stiffness, etc.) are changed significantly. As a result, the color of the specimen becomes dark brown and the specimen is very hard and brittle if thermal cured. Therefore, it is not so advisable to create parts by thermal curing.

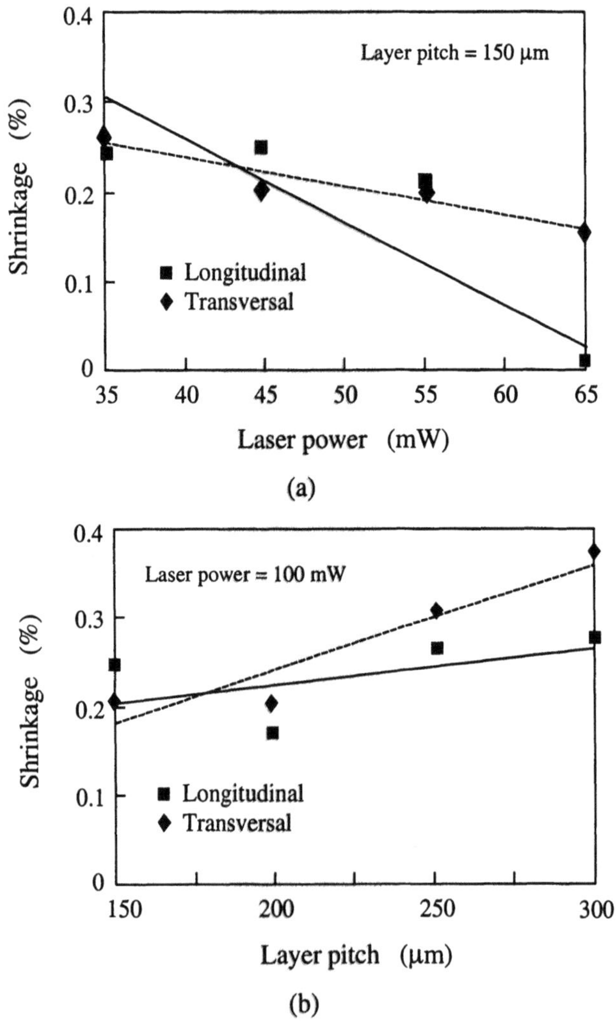

Fig. 3.9 Shrinkage vs. laser power and layer pitch after 2-hour UV post-curing.

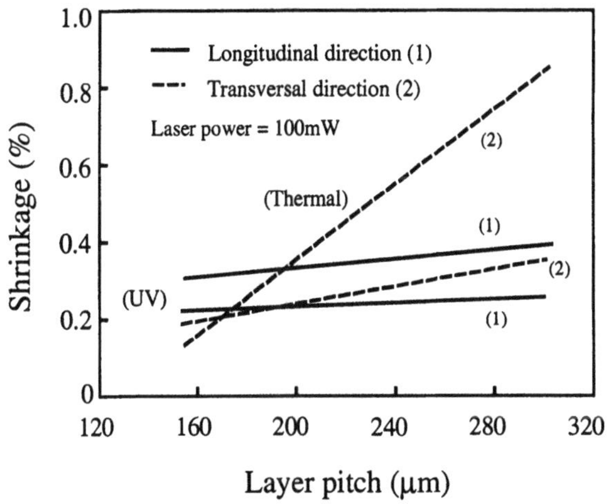

Fig. 3.10 Comparison of shrinkages from thermal and UV post-curing.

3.2 Analyses of degree of curing

This section discusses a detailed study on the degree of curing (or curability) of LL built part using Raman spectrum analysis at the micro-scale and differential scanning calorimeter (DSC) at the macro-scale. The analyses will help understand the complex behaviors of photo-polymerization occurring during the laser scanning process and the in-homogeneous structure as a result of laser fabrication process. Thus, the LL process used to build rapid prototypes can be better controlled by increasing the curability and reducing the resultant shrinkage and distortion.

3.2.1 Raman spectrum analysis

The Raman effect is a phenomenon observed in the scattering of light as it passes through a material. Raman scattering differs from other forms of light scattering as the photon suffers a change in frequency and a random alternation in phase. Fig. 3.11 shows a schematic diagram of a Laser-Raman Spectroscopic system. On passing through a material medium, the photons bounce off the molecules in an inelastic collision, and gain energy or lose energy. The change in energy is related to the shift in frequency by:

$$\Delta E = h\Delta \nu \tag{3.2}$$

where ΔE is change of energy; h, the Plank's constant; and $\Delta \nu$, change of frequency.

Fig. 3.12 shows a typical Raman spectrum of an acrylic-based photo-polymer. The x-axis is delta wavenumber (frequency/speed of light) and the y-axis is intensity of the scattered light. The delta wavenumbers of the peaks are consistent with repeated scattering [21,22] and the intensity of the Raman spectrum is dependent on the power of the incident laser. Polyacrylate polymers can be obtained from a variety of acrylic monomers, such as acrylic and methacrylic acids, their salts, esters, and amides, and the corresponding nitrides. For polymer resins, it is known that the peak on a Raman spectrum at a delta wavenumber spanning from 1600 cm^{-1} to 1670 cm^{-1} is a result of carbon double bond (C=C) present in the resin. To quantify the amount of monomer present, the amount of carbon double bond, as determined using the intensity of the peak occurring from 1600 cm^{-1} to 1670 cm^{-1}, was used for the evaluation of curing percentage.

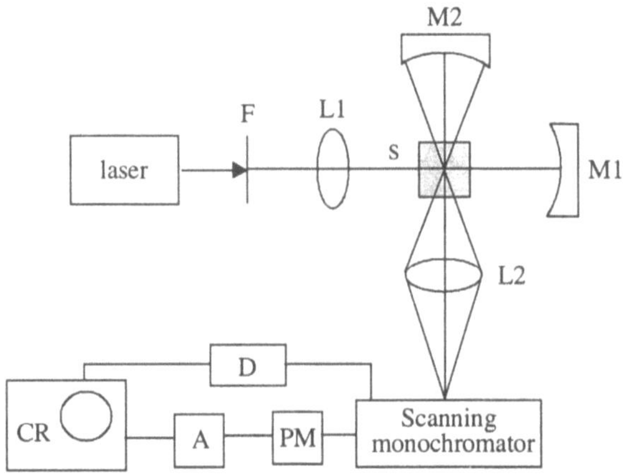

Fig.3.11 Schematic diagram of Laser-Raman Spectroscopic system.

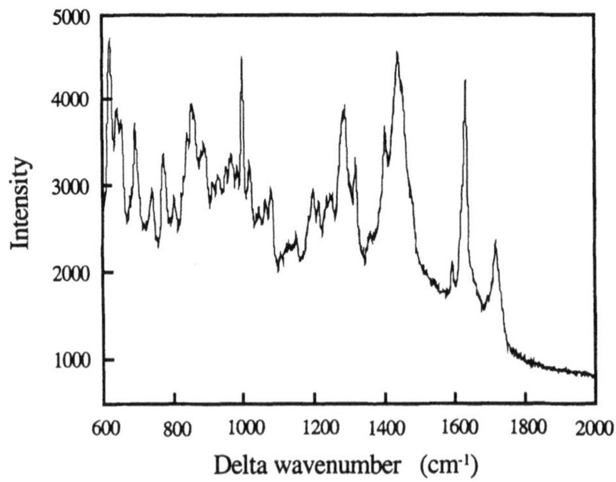

Fig. 3.12 A Raman spectrum example.

Assuming that any photon colliding with each C=C bond results in a contribution to the peak at delta wavenumber around 1633 cm^{-1}, we can take the intensity of peak at 1633 cm^{-1} to be a function of:

$$I_R = I\,P(N_{cdb})\,n \tag{3.3}$$

where I_R is the intensity of peak in the Raman spectrum; N_{cdb}, the number of carbon double bonds C=C in the specimen, $P(N_{cdb})$, the possibility of a light particle colliding a C=C bond, and n the number of light particles emitted by the Raman machine. Assuming that $P(N_{cdb})$ is a linear function to N_{cdb}, i.e.

$$P(N_{cdb}) = kN_{cdb} \tag{3.4}$$

then,

$$I_R \propto N_{cdb} \tag{3.5}$$

The percentage of curing is defined as the ratio of energy released to the total energy that can be released during polymerization. Assume that every reaction in the resin where a monomer links up with the polymer chain, i.e. the "breaking up of the carbon double bond", releases a fairly constant amount of energy and the intensity of the peak occurring at 1633 cm^{-1} to be of a linear relationship with the percentage of curing. This relationship can be used to determine the curing degree.

3.2.2 DSC analysis

As photo-curing is a highly exothermic process, heat is released by the resin during photo-curing and hence can be used to measure the amount of curing. The differential scanning calorimeter (DSC) is an instrument that measures the amount of heat released or absorbed in a reaction. The DSC was employed to measure the heat released during curing and hence the curing percentage. The curing percentage of the partially cured polymer line is defined by the ratio of heat released of the polymer line to the total heat released of the liquid resin. The following equation was used to determine the degree of curing of the built part:

$$\text{Degree of Curing (\%)} = (1 - H_{spc} / H_{liq})\,100\% \tag{3.6}$$

where H_{spc} and H_{liq} are the quantities of heat released from the laser-cured specimen and liquid resin respectively during DSC tests.

The measured degree of curing for the different specimens plotted against laser power and layer pitch is illustrated in Fig. 3.13. It is obvious from the plot that an increase in the laser exposure density will lead to a better cured structure for any particular layer pitch as it shows a higher degree of curing. This can be explained by the increase in the critical radius $L_w/2$ brought about by the increased penetration power of the laser beam. This results in the formation of a successive series of single-line cures having a wider width and also a greater cured depth. This leads to a more compact cured structure with less retained uncured resin. It can also be observed that the degree of curing increases with a decrease in the layer pitch. This is attributed to the cure depth of each scanned line. For a particular laser power, a definite amount of cure depth can be obtained. In practice, the layer pitch is often set smaller than the cure depth for a particular laser power. This is to provide for some over-curing of the previously scanned layer to occur, in order to allow the bonding of successive layers. Therefore, a decrease in the layer pitch with constant laser exposure density will increase the amount of over-curing of the previous layer leading to a more compact structure.

Since the DSC only reveals the average heat release of the UV scanned line, Raman spectroscopy can therefore be used to find the curing percentage at different locations of the cross-sectional area. This is achieved by first calibrating the Raman spectra to different curing percentages of the polymer; followed by comparing the calibrated results with spectra obtained at different locations of the cross-sectional profile of the polymer line.

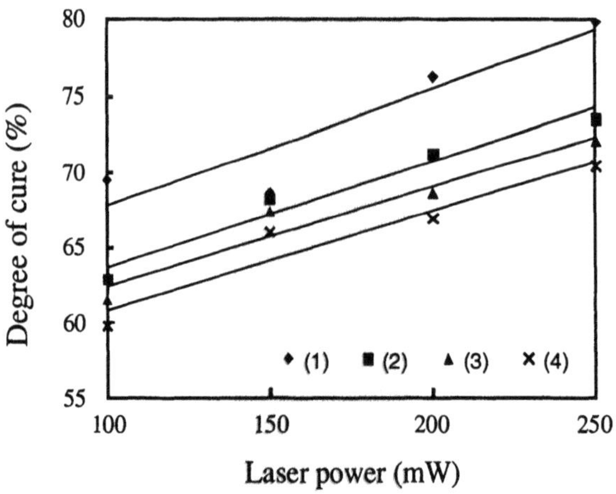

Fig. 3.13 Variation of degree of curing with laser power and layer pitch, where (1) layer pitch of 150 μm, (2) 200 μm, (3) 250 μm, and (4) 300 μm, respectively.

3.2.3 Processes

The SONY JSC-2000 LL machine was used to produce the specimens. Single polymer lines were created in a room lit only by monochromatic light and protected from UV exposure. The scanning velocity of the UV laser was 800mm/s and the power used was 250mW. Single polymer lines of 5cm long were formed by the UV laser scanning on a vat of resin. The cross section of the single polymer line is shown in Fig. 3.14. It is categorized into different areas for analysis.

The uncured liquid polymer was scanned using the Raman machine to obtain its spectrum. To get the fully cured specimen, 1-inch cubes produced using the LL machine were placed in the UV chamber for one hour to ensure 100% curing. Raman spectra of the fully-cured specimens were then obtained. For the DSC experiments, 2.0 mg of the liquid photo-polymer was placed in aluminum hermetic pans and exposed for 2s, 4s, 6s, 8s and 12s. Energy released was recorded using a DSC. Energy released during this initial exposure was compared against the total energy. The total energy is the summation of the initial energy and the residual energy released as the resin will undergo a second final exposure.

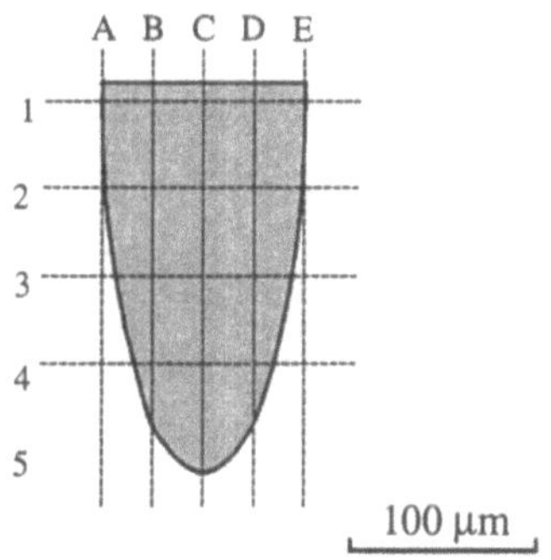

Fig. 3.14 Segregation of cross-section of a single scanned line.

DSC measurements were carried out on the samples using an AT Instruments, DSC2920 DSC machine. To obtain accurate results from the DSC, measurements have to be done on samples of the specimens immediately after laser fabrication in order to minimize environmental influences such as temperature and sunlight. A thin slice from each test specimen was weighed and encapsulated inside thin aluminum pans to form a pellet. The pellet, along with an empty reference pellet, was placed on top of two separate heaters inside the DSC cell. The heating cycle was then input into the DSC instrument. The temperature of the cell was raised to 300°C from room temperature at a rate of 20°C/min. The heat flow across the thin sample specimen was plotted against temperature and the amount of heat released from the sample was calculated by integrating the area below the heat flow graph. This value was automatically normalized with the weight of the specimen to give the amount of heat released by a gram of the specimen.

3.2.4 Analyses

Raman Spectra of Uncured (liquid) and Fully Cured Resin

Fig. 3.15 shows the typical Raman spectra obtained from the 0%- and 100%-cured polymer respectively. The representative change of the peaks at delta wavenumber 1633 cm^{-1} between the 2 spectra is evident. A prominent peak occurring at delta wavenumber 1444 cm^{-1} would be used as a standard for comparison. Estimating each peak in the Raman spectrum as a Lorentzian curve; intensities and areas under peaks at delta wavenumber of 1444 cm^{-1} and 1633 cm^{-1} can be obtained. Table 3.1 shows the values obtained from the Raman spectrum on the resin. In order to compare spectra from different scans, the values of intensity and area under peak obtained at delta wavenumber occurring around 1444 cm^{-1} is being used for comparison. This is also known as the internal standard method [22]. It is observed that the peak occurring around delta wavenumber 1633 cm^{-1} undergoes changes during the polymerization process. There is strong evidence in the results to indicate a reduction in both intensity and area for the peak. This is in accordance with results from most polymers where the peak representing the C=C bond occurs around delta wavenumbers 1650 cm^{-1}.

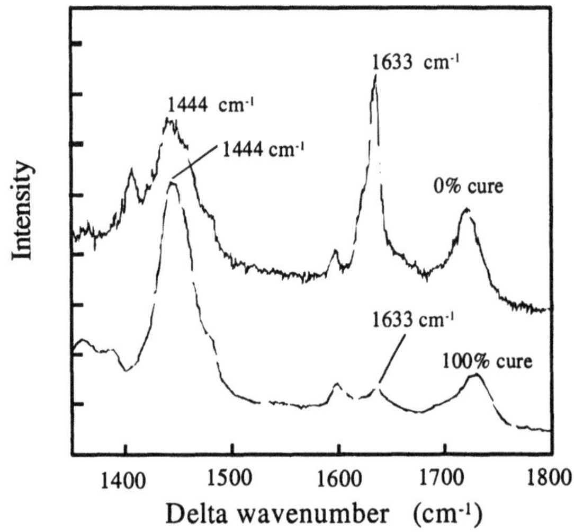

Fig. 3.15 Raman spectrum of 0%- and 100%-cured polymer [24].

Table 3.1 Values taken from a Raman Spectrum.

Peak no	Delta wave-number	Intensity of peak	Area under peak
(1)	1444	143.6713	11928.41
(2)	1633	189.5606	5290.175
Ratio of (2)/(1)	-	1.3194	0.4435

Raman Spectra of Various Regions of Single Scanned Lines

For the experiment, the various regions in the profile were segregated to investigate the degree of curing within a single scanned line produced. To obtain the percentage of curing at various regions in the profile of a single scanned line, the Raman spectra were analyzed based on the intensities and areas under peaks. Fig. 3.16 is the Raman spectra for position C_1 (i.e. the top position, see Fig. 3.14) at the single scanned line before and after post-cure for delta wavenumber 1633 cm^{-1} and 1444 cm^{-1}, respectively. For the ratio of intensities along various points on the center vertical line, it can be seen that the data collected do follow a trend, whereas the ratio of the areas along various points on the center vertical line do not follow as closely to one another.

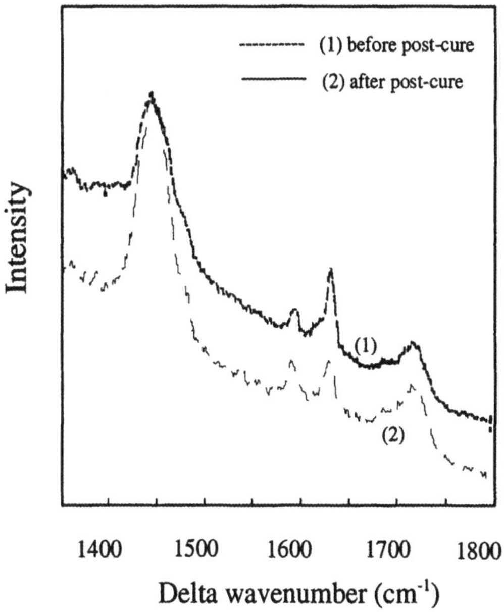

Fig. 3.16 Raman spectrum for single scan line at top position, C_1.

Curing Degree

Table 3.2 is the intensity ratio and its converted average curing percentage of various points on the scan line before post-cure. Similarly, the data for post-cured scan line is obtained and shown in Table 3.3. After post cure, the average percentage of curing ranges from 87.6% at the bottom to 96.4% at the top. The results based on intensities of the peaks show more consistency as compared with the results based on areas under peaks. Experiments showed that it was not necessary that the least cured area lied at the lowest end of the single scanned line. Also, the top surface of the scanned lines did not have 100% cured polymer as expected. The lowest curing percentage came as low as 62%. Comparing the curing percentages at green state and after post curing, the curing percentage change was about 30%. The maximum difference in the curing percentage was about 16% (based on intensity). However, given the quantifying accuracy of the Raman technique to be 10%, there is no evidence that much difference appears in curing percentages at green state.

Table 3.2 Intensity ratio and average curing percentage on the single scan profile (before post-cure).

Intensity ratio / (Curing degree %)	Position A	Position B	Position C	Position D	Position E
1	0.485/ (75.02)	0.500/ (74.73)	0.473/ (73.11)	0.500/ (74.73)	0.485/ (75.02)
2	0.448/ (73.62)	0.359/ (78.86)	0.420/ (76.76)	0.359/ (78.86)	0.448/ (73.61)
3		0.416/ (75.88)	0.386/ (79.44)	0.416/ (75.88)	
4		0.459/ (71.66)	0.347/ (78.41)	0.459/ (71.66)	
5			0.547/ (62.26)		

Table 3.3 Intensity ratio and average curing percentage on the single scan profile (after post-cure).

Intensity ratio / (Curing degree %)	Position A	Position B	Position C	Position D	Position E
1	0.167/ (96.4)	0.169/ (95.9)	0.165/ (94.7)	0.169/ (95.9)	0.167/ (96.4)
2	0.281/ (86.4)	0.174/ (94.9)	0.205/ (92.6)	0.173/ (94.9)	0.281/ (86.4)
3		0.204/ (93.05)	0.184/ (94.6)	0.204/ (93.1)	
4		0.199/ (89.6)	0.166/ (95.1)	0.199/ (89.6)	
5			0.252/ (87.6)		

Fig. 3.17 shows a DSC scan for a initial 8 seconds (s) UV exposure followed by a 120s UV final exposure. The initial heat released at 8s is 116.38 J/g, and 49.77 J/g after the final 120s exposure to achieve a total of 92.46% curing (see Table 3.4). There is also evidence that between the exposure time of 4s to 12s, there is a linear relationship between the curing percentage and time of UV exposure.

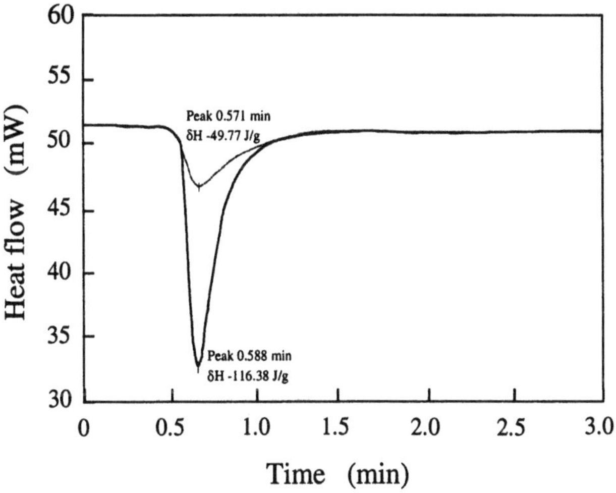

Fig. 3.17 DSC scan for initial 8s UV exposure followed by 120s final exposure.

Table 3.4 Results of DSC reading.

sample	mass (mg)	initial-exposure-time(s)	initial-energy-release(J/g)	final energy release(J/g)	total energy release	% cure after initial exposure
1	1.953	2	12.12	151.53	163.65	7.406
2	2.078	4	71.31	77.82	149.13	47.817
3	2.004	6	95.51	69.26	164.77	57.966
4	2.048	8	116.38	49.77	166.15	70.045
5	2.032	12	145.8	11.89*	157.69	92.460*

(final exposure time is 180s compared to others of 120s)*

3.3 Distortion analysis by Moiré method

3.3.1 Principles

Besides shrinkage, distortion (Fig.3.18) is another major problem for LL process and material. The Moiré method is a versatile set of techniques for in-plane and out-of-plane deformation measurement, topographic contouring, and slope and curvature measurement. The Moiré method for strain analysis can be used to determine the surface distortions on LL-fabricated specimens and to produce experimental results for analysis. The procedures and experimental set-up using the Moiré method will be described in the later section.

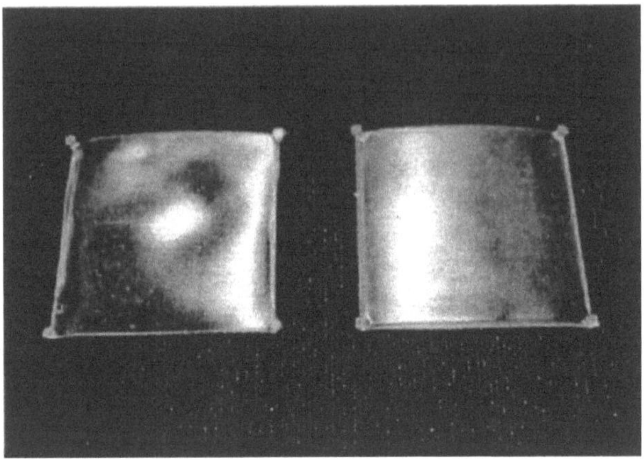

Fig. 3.18 Curl distortion encountered by thin-layered specimen.

3.3.1.1 Formation of Moiré fringes

In analyzing the fringe patterns formed using Moiré analysis, two methods are widely documented, namely, the *geometrical* and the *displacement*-field methods. In the following, the displacement-field method is used to analyze the fringe patterns obtained. Moiré fringe patterns can be formed using arrays consisting of a series of straight parallel lines, concentric circles, etc.

For works involving stress-strain investigations, arrays with straight opaque parallel lines and transparent inter-spaces are the most widely used. The pitch of the arrays or otherwise known as gratings, is available up to 1,000 lines/inch. In Moiré analysis, the sensitivity of the method is dependent upon the pitch of the gratings used [25]. The accuracy of the measurements improves with the usage of gratings with increasingly fine grids. To produce Moiré fringes, the superposition of two gratings (one to serve as the deformed specimen grating while the other the reference grating) by mechanical or optical means has to be carried out. Both specimen and reference gratings are identical when the body on which the specimen grating is attached or formed, is in the un-deformed state. This is critical, as any differences in the two gratings at this state, will produce inaccurate results.

The different techniques used in Moiré analysis vary, depending on how the specimen grating is generated. For in-plane Moiré [26] measurements, the specimen grating is formed by printing or bonding thin films carrying the grating onto the surface of the body under study, where it faithfully follows the deformation. For out-of-plane measurements (i.e. shadow Moiré method [27, 28]), the specimen grating is a shadow of the reference grating projected onto the specimen's surface.

Moiré fringe patterns are formed when light is transmitted through the reference grating onto the specimen grating. The direction transverse and parallel to the grating are termed as the primary and secondary directions, respectively. To measure the strains along any particular axis, the primary direction has to be aligned parallel to that axis. The fringes appear as a series of dark and bright bands parallel to the secondary direction of the gratings. Individual fringes appear when deformations encountered by the body in the primary direction causes an extension or contraction in the specimen grating by a magnitude equal to the pitch, *p*, relative to the reference grating. When a single directional grating is used, any deformation encountered by the body in the secondary direction does not produce any fringes. The change in length Δl, encountered at any point on the surface of the specimen, is given by:

$$\Delta l = n\,p \tag{3.7}$$

where *n* and *p* are the numbers of fringes encountered in going from the 0-order fringe to the point in note (gauge length, l_g) and the pitch of the non-deformed/reference grating, respectively.

Thus the engineering strain in this interval can be expressed as follows:

$$\varepsilon = \frac{np}{l_g - np} \quad \text{for tensile strains} \tag{3.8}$$

$$\varepsilon = \frac{-np}{l_g + np} \quad \text{for compressive strains} \tag{3.9}$$

3.3.1.2 Displacement-field method for Moiré fringe pattern analysis

Since the specimen grating follows the same displacement field as the deformed body, the Moiré fringe patterns can be related to a displacement field. This forms the basis of the displacement-field method for Moiré fringe pattern analysis. Fig. 3.19 illustrates the Moiré fringe pattern formed on the surface of a deformed specimen. In the un-deformed state, both specimen and reference gratings are identical in the sense that the lines on both gratings coincide with one another. For clarity of explanation, the grids of both the specimen and reference gratings are numbered. The grids on each of the two gratings with the same assigned number are assumed to coincide before deformation is introduced on the body. From the fringe pattern illustrated in the figure, the fringe that shadows the intersection points of grid lines with the same number is taken as the 0-order fringe. Fringes with the order, *n*, of 1, 2, 3, etc. are formed by the intersection of reference grid number 0 with specimen grids numbering 1, 2, 3 respectively. Since Moiré fringes are interpreted as the loci of points on the deformed surface of the body having the same displacement component, from Eq. (3.7), points covered by the nth-order fringe are displaced a distance equal to the value of *np*. To analyze the deformed surface, the surface is discretized into two successive series of lines parallel to the *x*- and *y*- axis. A displacement against distance from the 0-order fringe graph (i.e. displacement surface) is generated for each of line. From the displacement surfaces, the Cartesian components of strain (i.e. d*u*/d*x*, d*u*/d*y*, d*v*/d*y* and d*v*/d*x*) at any point on the deformed surface can be calculated from the gradient of the graph at that particular point.

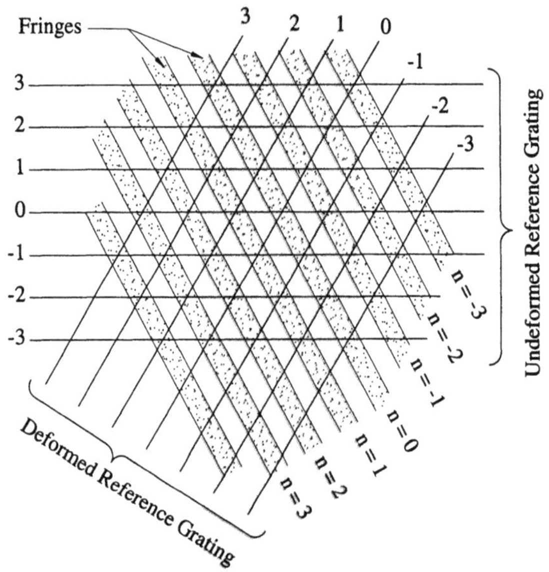

Fig. 3.19 Moiré fringe formation on surface of stressed specimen.

For cases where small values of strains are encountered, the strain components can be calculated using the following equation:

$$\varepsilon_{xx} = \frac{du}{dx}, \quad \varepsilon_{yy} = \frac{dv}{dy}, \quad \gamma_{xy} = \frac{dv}{dx} + \frac{du}{dy} \tag{3.10}$$

where *u*, *v* and *w* are the displacements measured in the *x*, *y* and *z* directions respectively.

3.3.2 Processes

To compare the surface deformation of pure and composite specimens, two sets of specimens can be generated. The test piece is a square with 80 mm sides and 4 mm thickness. After laser fabrication, the specimens are gently removed from the building platform and their surfaces are wiped with solvents using soft tissues to remove all uncured resin. This is to prepare a clean surface for the bonding of the grating film. The gratings chosen for this set of experiments are uni-directional and have a pitch of 0.01mm (100 lines/mm). The gratings are then bonded, film side down, onto the top surface of each specimen to form the specimen grating. This can be done using Cyano Acrylate glue, which will be left to cure for 24 hours. After curing, the backing of the film is carefully removed using a pair of tweezers. A fine tip marker can then be used to draw out a mesh of points spaced 10mm apart on the surface of the film to serve as reference points for the measurements to be carried out. Fig. 3.20 illustrates the set-up can be used to post-cure the specimens and to photograph the formation of the fringe patterns.

To form the reference grating, a piece of grating with the same pitch as the specimen grating was bonded onto a piece of thin, transparent non-reflective glass. In the set-up, the reference grating was aligned directly above the specimen by using an adjustable mount. Before the experiment could be started, the lines on the master grating have to be aligned parallel to those on the specimen grating. This was achieved by manually rotating the master grating relative to the specimen grating until no fringes could be observed through the gratings. The UV tubes were then lighted and the computer was set to capture an image of the specimen at an interval of five minutes. For the determination of in-plane strain calculations, the displacement-field method described in section 3.3.1.2 was used.

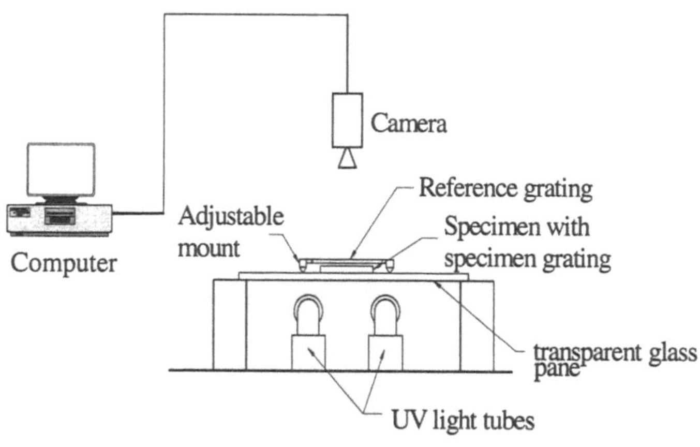

Fig. 3.20 Set-up for determination of surface distortions using the Moiré method [29].

3.3.3 Analyses

The fringes generated using Moiré gratings allow the shrinkage at any point on the surface of the specimens along both the scanning (longitudinal) and transverse directions to be determined. The amount of surface distortions encountered upon laser and post-curing can also be calculated.

Fig. 3.21 (a) and (b) show the Moiré fringe patterns formed for shrinkage measurement of SCR310-built specimens. The laser power setting was 150mW and the layer pitch was set at 150μm. The fringe patterns formed due to the deformation of the specimen surface are observed as black and white bands of light. Fig. 3.21 (a) shows that the fringe pattern is parallel to the laser scanning direction whereas from Fig. 3.21 (b), the fringe pattern is perpendicular to the laser scanning direction. Since fringe patterns are formed perpendicular to the direction at which shrinkage occurred, Fig. 3.21 (a) and (b) are used to measure the transverse and longitudinal shrinkages, respectively. By spacing the lines at equal intervals along the specimen length, the strain encountered by any point on the specimen surface can be obtained.

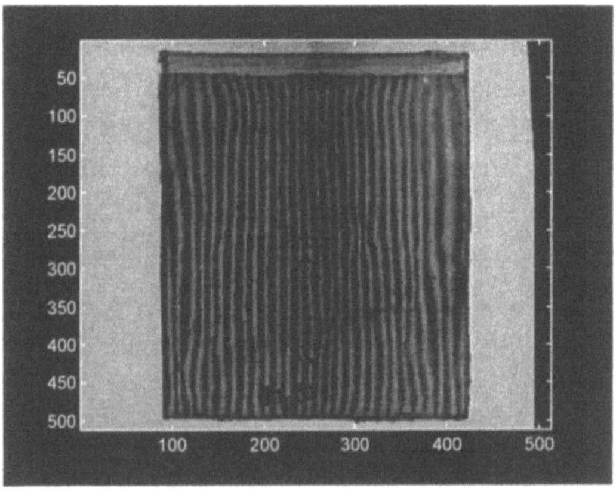

(a) Measurement of transverse shrinkage

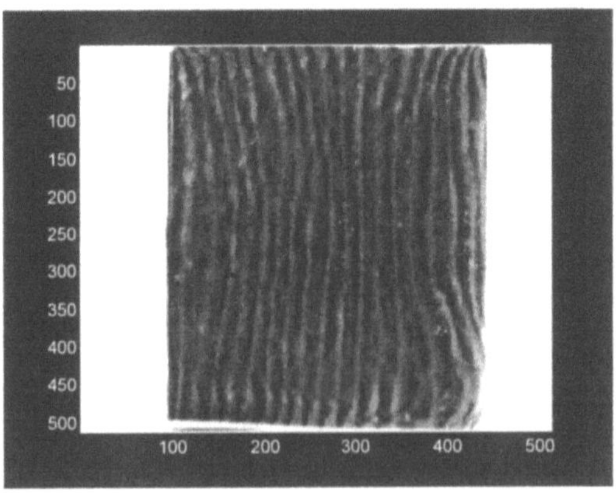

(b) Measurement of longitudinal shrinkage

Fig.3.21 Moiré fringe patterns for shrinkage strain measurements.

With the Moiré method, the specimens are subjected to free shrinkage as no physical constraints are imposed on them. The shrinkage encountered between two points on the surface of the specimen can be expressed in terms of a relative displacement between the two points. To calculate the amount of shrinkage encountered at any particular point on the surface of the specimen using the displacement field approach in Moiré fringe analysis, a centerline parallel to the fringe pattern has to be drawn on the specimen. Any fringe band coinciding with this centerline will be termed as the 0-order fringe. The amount of movement encountered by any particular point on the surface of the specimen relative to the center line drawn will be given by np (Eq. 3.7). The shrinkage strain at that particular point is then obtained as explained in section 3.3.1.1.

To compare and to study the fringe patterns presented in Fig. 3.21, imaginary lines are drawn on the center of each specimen, perpendicular and parallel to the fringe patterns. Comparing the fringe pattern as shown in Fig. 3.21 (a) and (b), it can be observed that more fringes are formed parallel to the longitudinal direction (Fig. 3.21 (a)) than perpendicular to it (Fig 3.21 (b)). The dark fringe forming at the center of the specimen as observed in both Fig. 3.21 (a) and (b) is taken as the 0-order fringe. From Fig. 3.21 (a), a point located on the left edge of the specimen along the imaginary line drawn perpendicular to the fringes is covered by a bright fringe, which is the 13.5-order fringe. Compared to a point on the same location in Fig. 3.21 (b), it was observed that the dark fringe encountered at that particular location is the 9th-order fringe. The longitudinal strain is calculated to be 0.23% while the transverse strain at the same point is found to be 0.34%. By observing Fig. 3.21 (b), the extrapolated values of longitudinal and transverse shrinkage strains obtained through measurement using a two-axis optical microscope were found to be 0.257% and 0.41%, respectively. Longitudinal and transverse shrinkage values obtained through measurement using the optical microscope were found to be larger in magnitudes than those obtained through Moiré analysis. However, the trends of shrinkage behavior exhibited by the specimens are the same for both measurement techniques. The difference in the measured values may be due to the fact that the specimens warp and curl slightly upon post-curing mainly due to the uneven amounts of shrinkage encountered within their structures. The optical microscope is only capable of measuring the projected and not the actual length of the top surface of the specimen. Warping in the specimens causes the measured projected lengths to be much shorter compared to their actual value thereby giving higher values of shrinkage as observed above.

The Moiré strain analysis method is found to be accurate in giving actual strain readings over an entire deformed surface since measurements are carried out directly on the specimen surface. Its only limitation is in the fineness of the gratings used. For the experiments carried out, the gratings used had a pitch of 0.01mm and this was deemed suitable. The Moiré strain results can be used to verify the results given by FEM simulation as described in reference [29].

3.4 Summary

Laser-lithography (LL) is a new and advanced manufacturing process for the generation of complex 3-D models within a much shorter time and at a lower cost. Despite its great potential in the industry, it is plagued by accuracy issues arising from the curability of the photo-curable resin and its weak mechanical properties. Although post-curing is necessary to give the prototype its ultimate strength, thereby, increasing its durability, the significant amount of shrinkage encountered in the post-curing process is one of the major source of errors in the LL process. It is therefore important to increase the degree of curing at the green state and as a result to minimize the post-cure shrinkage as much as possible. The curability and mechanical properties of rapid prototypes built by LL process have been described in this chapter. Differential scanning calorimeter and Raman spectrum have been used to analyze the degree of curing of the photopolymer after laser scanning. The curability and mechanical properties of a LL-built part are found to be a function of layer pitch and laser exposure density. From tensile tests, it is known that the elastic modulus, ultimate tensile strength and elongation to fracture of the specimens are also functions of laser exposure density and layer pitch. By increasing the laser exposure density and decreasing the layer pitch, the curing percentage of the photo-polymer can be increased, leading to parts built with better mechanical strengths.

3.5 References

1 G.R. Tryson, and A.R. Shultz, *J. Polym. Sci., Polym. Phys. Ed.*, Vol. 17 (1979),125.
2. C.E. Hoyle and J.F. Kinstle (Eds.), *ACS Symp.* Ser.(1990), 417.
3. A.J. Lightman, and R.E Anderson, *SAE Trans., J. of Aerospace*, Sec.1 (1991), 303.
4. C. Seigne, J.Y. Jezequel, P. Schaeffer, and J.C. Andre, *Proc. of 3rd Inter.Conf. on Rapid Prototyping*, Dayton, Ohio, (1992), 131.
5. J. Paczkowski and D.C. Neckers, *J. Polym. Sci.*, Polym. Chem., Vol.31(3) (1993), 841.
6. C. Decker and K. Moussa, *J. Polym. Sci.*, Polym. Lett. Ed., Vol.27 (1989), 347.
7. C. Decker and K. Moussa, *J. Coatings Techn.*, Vol. 65 (819), (1993), 49.
8. P. Karrer, S. Corbel, J.C. Andre, and D.J. Lougnot, *J. Polym. Sci.*, Polym. Chem., Vol. 30 (13), (1992), 2715.
9. J.Y.H. Fuh, Y.S. Choo, A.Y.C. Nee, et al., *Materials & Design*, Vol 16(1) (1995), 23.
10. C.M. Cheah, A.Y.C. Nee, J.Y.H. Fuh, et al., *J. Mater. Proc. Techn.*, Vol.67 (1997), 41.
11. C.M. Cheah, J.Y.H. Fuh, A.Y.C. Nee, et al., *J. Mater. Proc. Techn.*, Vol.(67) (1997), 46.
12 W. Kunze, *Technical Report from Perkin-Elmer*, Uberlinen, FRG and Bernhard Stapp Siemens AG, Erlangen, Germany (1994).

13. K. Arita, A. Ikeda, et al., *Proc. of 1991 Inter. Conf. on Ind. Elec. Control & Instru.*, IECON '91, (1991), 74.
14. P.F Jacobs, *Rapid Prototyping & Manufacturing : Fundamentals of Stereolithography*, Society of Manufacturing Engineers (SME), Dearborn, MI, (1992).
15. P.T. Weissman, B.A. Bolan, *Proc. of 3rd Inter. Conf. on RP*, Dayton, Ohio, (1992), 103.
16. L. Lu, J.Y.H. Fuh, A.Y.C. Nee, et al., *Mate. Res. Bul.*, (1995),1564.
17. W. Nau, R.L. Dooley, and A.A. Ogal, *Proc. of 5th Inter. Conf. on RP*, Dayton, Ohio, (1994), 121.
18. S. Iwanaga, and Ohkawa, *Proc. of 3rd Inter. Conf. on RP*, Dayton, Ohio, (1992), 77.
19. *SCR-310*, the brand name of UV curable resin from Nippon Synthetic Rubber, Ltd., Japan (1998).
20. J.Y.H. Fuh, Y.S. Choo, et al, *J. Mater. Proc. Techn.* ,Vol.53 (1997) 887.
21 S. Bhagavantam, *Scattering of Light and the Raman Effect,* Chemical Publishing Co, Inc., 1942.
22 S. K. Freeman, *Applications of Laser Raman Spectroscopy,* John & Wiley, N.Y., (1974).
23. G.R. Tryson and A.R. Shultz, *J. Polym. Sci.*, Polym. Phys. Ed., Vol. 17 (1979), 2059.
24. J.Y.H. Fuh, L. Lu,, et al., *J. Mater. Proc. Techn.*, 89-90 (1999), 211.
25. A. Asundi, *Optics and Lasers in Eng.*, Vol. 11 (4), (1989).
26. A. Asundii, *Photomech. and Speckle Metrology II*, SPIE, Vol. 1554, (1991).
27. A. Asundi, C.M. Wong, *Exp. Tech.*, Vol. 15 (1), (1990), 44.
28. A. Asundi, *NDT Inter.*, Vol. 24(1), (1991), 33.
29. C.M. Cheah, *PhD dissertation*, National University of Singapore (2001).

Chapter 4

IMPROVEMENTS OF MECHANICAL PROPERTIES BY REINFORCEMENTS

4.1 Introduction

Since the wider adoption of LL process in industry, much research has been carried out to develop prototypes with higher strength and accuracy [1-4]. The nature and origin of distortion in the LL-built part has been well studied and documented [5-8]. Recent R&D efforts in rapid prototyping have focused on the development of techniques for the fabrication of functional and semi-functional parts based on metal composites utilizing direct laser sintering techniques [9] or through metal infiltration processes [10,11] on the fabricated rapid prototype. The purpose of most reinforcements, such as glass fiber and ceramic power, is to increase the tensile strength and hardness. Work has also been carried out to improve the strength of LL parts by embedding continuous fiber tows into the fabricated part [12].

Although enhancement of mechanical properties can be achieved with these reinforcements, re-circulation of these mixtures may become a problem to the operator. As the mechanical properties of a composite part is highly affected by the concentration, shape, distribution and properties of the additives, many studies have been carried out to determine an additive that is suitable for the purpose. Current commercial production of composite parts is either by lamination, casting or molding. In the case of composites built by the LL process, besides being capable of improving the mechanical properties of the laser-fabricated parts, the choice of the composite additive depends heavily on its ease to be processed and incorporated into the existing LL machine set-up. The following sections present the principle and application of fibers and AEROSIL powders as reinforcement in photopolymer to improve the mechanical properties of LL-built parts.

4.2 Fiber-reinforced photo-polymer

Dimensional accuracy and strength of parts created using the LL method have been a major problem, especially in the fabrication of thin-wall hollow structures. The weak mechanical properties and high volume of shrinkage associated with the photo-polymer cause the part to warp and distort during laser fabrication. As mentioned earlier, these inaccuracies are augmented during the post-processing stage. The poor strength and dimensional accuracy limit the applicability of the final parts, especially for use as functional or semi-functional prototypes [13].

Reinforcement in composite in the form of continuous and chopped fibers are very popular in industry. Reinforcement in the powder form have also been successfully used. For example, ceramic powder has been added to the polymer and used in rapid molding methods. Different kinds of reinforcement, ranging from powders (e.g. ceramic, glass) to fibers (e.g. carbon, glass), have been tried and tested [14]. It is found that both the ceramic powders and carbon fibers are unsuitable due to their opacity to UV light, causing delamination in the built specimens. However, photo-polymer mixed with glass fibers and powders have yielded promising results showing improvements in both the mechanical properties and the dimensional accuracy of the specimens.

The use of continuous fibers requires the development of a separate system to lay the fibers onto the surface of the photopolymer [12]. It is difficult to control the laying of the fibers and achieve a uniform spread over the surface of each layer. Furthermore, having to lay the fibers before laser scanning, will definitely lengthen the time required to build the part. Parts built using continuous fibers have anisothropic properties [15], improving the mechanical properties only in the lay direction of the fibers. An alternative to the use of fibers in strengthening a part is by the use of short fibers or spherical particles. The mechanical properties of composites formed using short fibers are isotropic due to their random distribution [16], giving improvements in strength in all directions. The usage of these additives does not require any major modifications to the LL system. The reinforcement can be pre-mixed with the LL resin and this mixture can then be coated into a thin layer in the same manner as the pure resin. Where necessary, the reinforcement can be kept suspended in the mixture through the use of an agitator attached to the bottom of the vat.

4.2.1 Basic principles and theory

The mechanical behavior and analysis of composites with discontinuous, or short fiber reinforcement is generally more complex compared to continuous aligned fiber reinforced composites. Complexity in the analysis of the short fiber reinforced composites arises from the end effects encountered in using short fibers and also from the fact that it is almost impossible for all the short fiber to be of equal lengths. Further to this, the orientation of the short fibers within the composite is at random. To achieve an accurate analysis in the modeling of short fiber reinforced composites, considerations made to determine the distribution of the fiber length (or aspect ratio) and also their orientations are essential.

Since the achievable fiber volume fraction V_f is always lower in short fiber reinforced composites compared to the continuous fiber reinforced composites, the short fiber reinforced ones will never exhibit the stiffness and strengths of the continuous fiber composites. The short lengths of the fibers also result in the properties of the matrix material and the fiber/matrix interface to have a higher influence on the properties of the composite. It has been demonstrated that the stiffness of the short fiber composites is highly dependent upon the length or aspect ratio of the short fiber [17,18]. The usage of fibers with high aspect ratios will cause the mechanical properties of the short fiber composite to approach the values exhibited by the continuous fiber composite. It has also been shown that that short fiber composite fractures have a much lower longitudinal strain value, compared to continuous long fiber composites [19-21]. This may be due to the presence of a stress concentration factor at the fiber ends as demonstrated in the results of a finite element analysis conducted by Barker *et al.* [20]. The stress concentration factor becomes a constant at when the fibers appear sufficiently long (high aspect ratios). Works conducted by Chen [19] and Riley [21] also show that the strengths of the short fiber reinforce systems reaching a plateau strength with high fiber aspect ratios.

In studying the mechanics of short fiber reinforced composites, the mechanics of the continuous fiber composites should be considered. The basic mechanics of continuous reinforced composite has been derived from the simple concepts of the "rule of mixtures" [22]. In applying the equation, perfect bonding between the reinforcement and matrix material is assumed. The structure of the continuous reinforced composite is also considered as an array of straight aligned fibers embedded within the matrix material. The rule of mixtures for the longitudinal stiffness of the composite is given by:

$$E_c = E_f V_f + E_m V_m \tag{4.1}$$

where E represents the modulus of elasticity and V the volume fraction. The suffices f, m, and c refer to the fiber, matrix, and composite, respectively.

This simple model predicts the linear variation of the longitudinal modulus with fiber volume fraction very well [22]. It can be observed from Eq. 4.1 that if the stiffness of the fibers is very much larger than the matrix material, or at high fiber fractions, the overall stiffness of the composite is less influenced by the stiffness of the matrix material. The same can be applied to the fiber arrangement. For analyzing the strengths of the composite, due to the greater influence of the interfacial interactions between fiber and matrix, the usage of Eq. (4.1) is not as suitable.

For the short fibers to reinforce the matrix, perfect bonding should be achieved to allow effective stress transfer between the fiber-matrix interface. An increase in the stiffness of the composite is obtained due to the constraining effect imposed by the fiber on the matrix. However, due to imperfections in the interface of the fiber and matrix material encountered in real systems, a limit is imposed on the maximum stiffness achievable by the composite. Other factors that directly affects the stiffness of the short fiber reinforced composite includes the fiber volume fraction, aspect ratio and orientation.

4.2.1.1 Modeling of short fiber composite [22]

There are two well-accepted simply models for stress transfer based on assumptions made regarding the properties of the matrix material. These models are the Cox [23] model and the Kelly-Tyson [24] model. The Cox model will be considered in this particular case. In the Cox model, the representative element of the composite is taken to be a pair of perfectly bonded concentric cylinders with the inner and outer cylinders representing the short fiber and matrix respectively, as illustrated in Fig. 4.1. The Cox model assumes both the matrix material and the fibers to be linearly elastic.

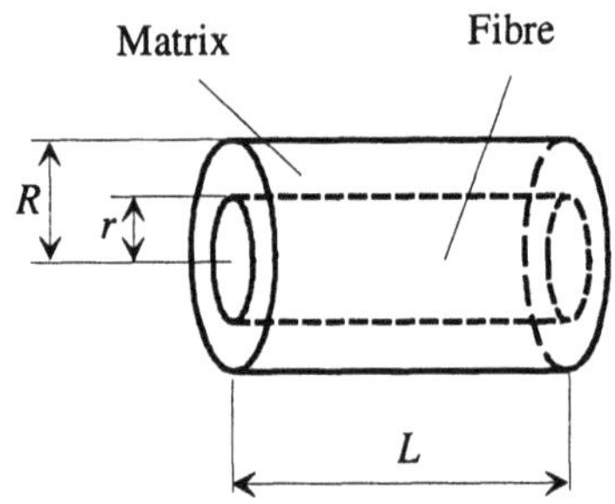

Fig. 4.1 A representative element based on Cox model.

In the representative element, the volume fraction of the fiber, V_f, is achieved by adjusting the ratio of (r/R). It is also assumed that stress transfer does not take place across the free ends of the representative element.

Consider a straight fiber of length L embedded in a solid matrix as illustrated in Fig. 4.1. In the Cox model, the rate of load transfer is proportional to the difference between the axial displacement, u, at a point in the fiber and the axial displacement, v, of the matrix at the same point in the absence of the fiber. The assumption of Cox model leads to the following equation:

$$\frac{dP}{dx} = H\ (u-v) \tag{4.2}$$

where P is the fiber axial load and H is a constant. By differentiating Eq. (4.2), the following equation is obtained:

$$\frac{d^2P}{dx^2} = H\left(\frac{du}{dx} - \frac{dv}{dx}\right)$$

$$= H\left(\frac{P}{A_f E_f} - \varepsilon\right) \tag{4.3}$$

where du/dx = $P/(E_f A_f)$ and dv/dx = ε. A_f and E_f represents the cross sectional area of the fiber and modulus of elasticity of the fiber respectively. By applying boundary condition values of $P = 0$ at $x = 0$ and L and solving Eq. (4.3), the fiber stress σ, and the interfacial shear stress τ (Fig. 4.2) can be obtained from the following:

$$\sigma_f = \frac{P}{A_f} = E_f\,\varepsilon\left(1 - \frac{\cosh\beta(L/2-x)}{\cosh\,(\beta L/2)}\right) \tag{4.4}$$

$$\tau = E_f\,\varepsilon\sqrt{\left(\frac{G_m}{2E_f \ln(R/r)}\right)}\frac{\sinh\beta(L/2-x)}{\cosh\,(\beta L/2)} \tag{4.5}$$

where

$$\beta = \sqrt{(H/E_f A_f)} \quad \text{and} \quad H = \frac{2\pi\, G_m}{\ln\,(R/r)}$$

G_m represents the shear modulus of the matrix material.

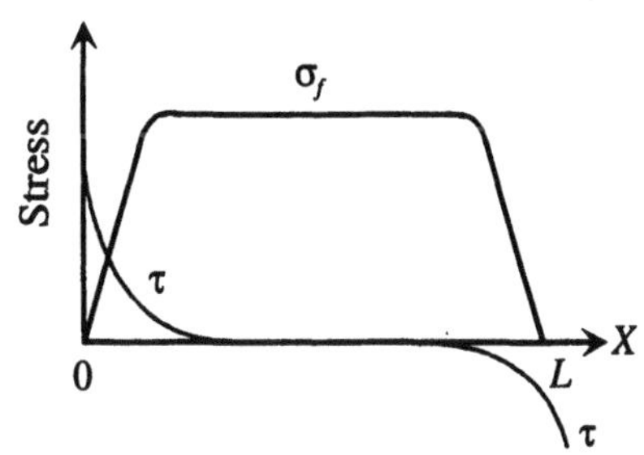

Fig. 4.2. Axial and shear stress distribution along a fiber.

4.2.1.2 Young's modulus and strength

To predict the stiffness of a short fiber reinforced composite with random fiber orientation, the "rule of mixtures" equation has to be modified with the inclusion of both the fiber orientation efficiency factor, η_o and the fiber length efficiency factor, η_l. The resulting equation for predicting the modulus of elasticity of the composite after the inclusion of both the efficiency factors takes the form of:

$$E_c = \eta_o \eta_l E_f V_f + E_m V_m \tag{4.6}$$

where

$$\eta_o = \sum a_n \cos^4 \theta \tag{4.7}$$

and

$$\eta_l = 1 - \frac{\tanh(\beta L/2)}{\beta L/2} \tag{4.8}$$

The fiber orientation efficiency factor, η_o, is based on Kenchel's work [25]; while the fiber length efficiency factor, η_l, is based on the Cox model [23].

The orientation of the fibers can be statically obtained based on the principle illustrated in Fig. 4.3. The major and minor axes of the ellipses formed by the fiber cross-sections, as illustrated in Fig. 4.3(a), were measured from printouts of images taken from a microscope. The fiber orientation distribution factor was calculated from Eq. (4.7). Based on the distribution of the fiber length, the average fiber length can be calculated using Eq. (4.8).

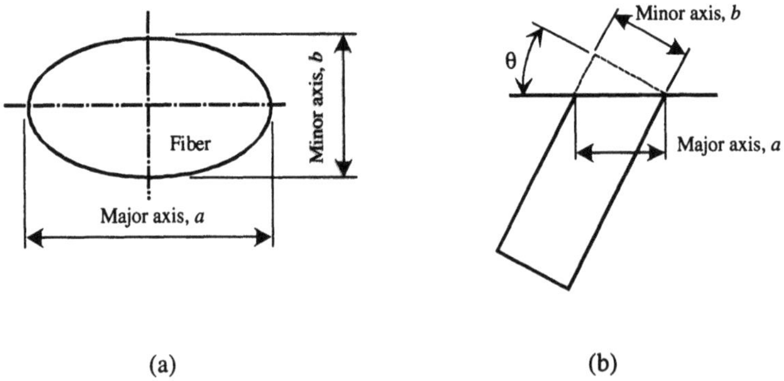

Fig. 4.3 Schematic of (a) cross-sectional profile of a single short fiber as observed from the top, (b) side view of fiber embedded within matrix

Two angles measured relative to the *z* and *y axis* are required to define the orientation of a randomly oriented fiber embedded within the matrix in three-dimensional space. The θ and ϕ are defined as the angles measured relative to the longitudinal axis (*z* axis) and the transverse axis (*y* axis) respectively as illustrated in Fig. 4.4. In this analysis, it is assumed that the short fibers (average length of 1.6mm) are oriented in a two-dimensional space. Thus, only one angle is needed to define its orientation and this is done using the angle θ.

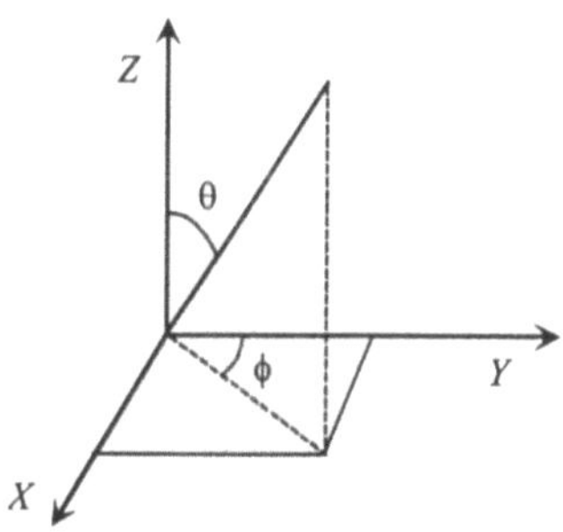

Fig. 4.4. Orientation of fiber in a three-dimensional space.

Fig. 4.5 illustrates the cross-sectional view of a molded and LL-built fiber reinforced SCR310 specimen. The gray elliptical spots at the top and top right-hand corner are cavities formed by entrapped air bubbles. Most of the fibers were observed to have an orientation in the plane of the specimen. These fibers strengthen the specimen in that direction. The major and minor axes of the ellipses formed by the fibers have to be measured from some randomly selected optical micrographs taken at different cross-sections along the length of the specimen.

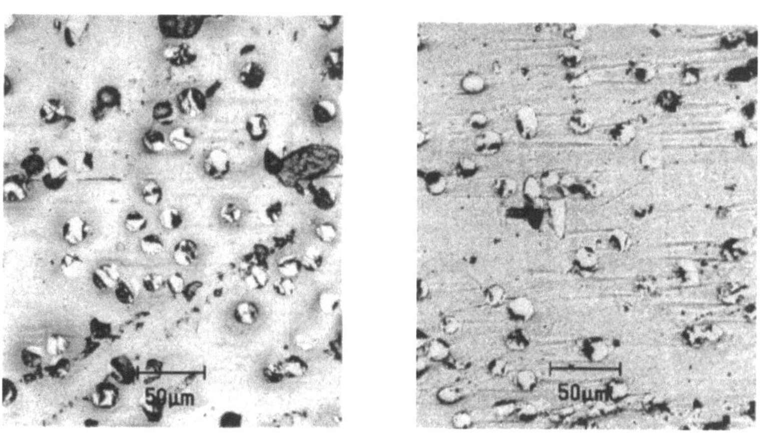

(a) (b)

Fig. 4.5 Microstructure of cross-section: (a) a molded fiber-reinforced SCR310 specimen, and (b) LL fabricated fiber reinforced, with 20% concentration by volume of fibers

Based on micro-structural observations, the frequency versus angle of inclination, and the frequency versus length of fiber are plotted, which can be used for the evaluation of η_0 and η_l. The distributions of the orientation and length are given in Fig. 4.6 and 4.7.

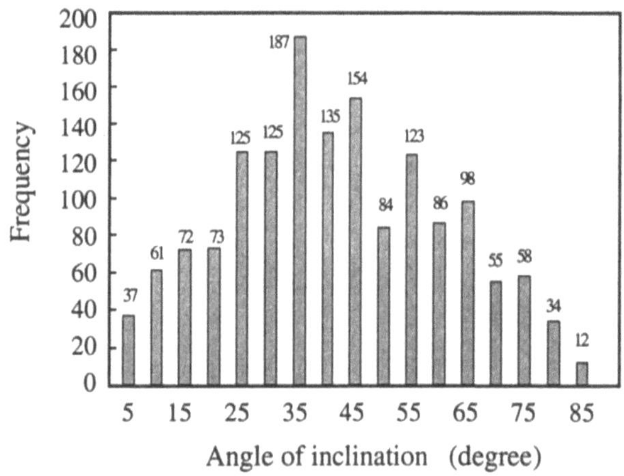

Fig. 4.6 Fiber orientation distribution.

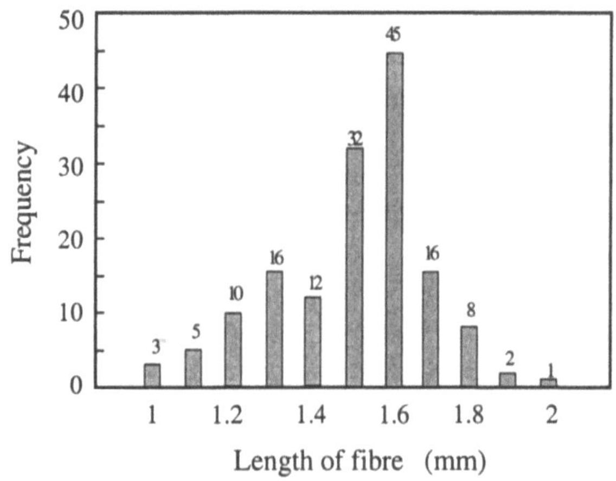

Fig. 4.7 Fiber length distribution.

To perform an analysis on the strength of short fiber reinforced composites, the original strength predictions developed by Kelly and his co-workers [24,26,27] were considered. The ultimate axial strength expression suggested by Kelly *et al.* is :

$$\sigma_{cu} = \sigma_{fu} V_f + \sigma'_{mu} (1 - V_f) \tag{4.9}$$

where σ_{cu} and σ_{fu} represent the ultimate tensile strengths of the composite and the fiber, respectively; σ_{fu} is identical with the fracture strength of brittle fibers; and σ'_{mu} denotes the stress in the matrix at the failure strain of the composite.

In deriving Eq. (4.9) for continuous reinforced composites, the applied load on the composite is assumed to be distributed between both the fibers and the matrix, and a uniform tensile strain relation occurs along the axial direction. When the reinforced fibers are short or discontinuous, Eq. (4.9) is no longer appropriate due to the invalidity of the iso-strain condition. The amount of strain difference between both fiber and matrix near the fiber end generates shear stresses along the axis of the fiber. Tension or compression arises in the fiber resulting from the presence of the shear forces. For short fibers, the applied load can only be dispersed among the fibers through this method of stress transfer.

To predict the strength of short fiber reinforced composites, modifications to Eq. (4.9) to include both the fiber length and orientation efficiency factors have to be carried out. Eq. (4. 9) becomes:

$$\sigma_{cu} = \eta_o \eta_l \sigma_{fu} V_f + \sigma'_{mu} (1 - V_f) \tag{4.10}$$

The efficiency factors, η_o and η_l, are the ones as given by Eq. (4.7) and (4.8).

4.3 AEROSIL-mixed photo-polymer

Present LL rapid prototyping methods require supports to be added to all overhanging parts. Without supports, it may not be possible or easy to build some models. The need for support arises due to gravitational effect and shrinkage encountered during modeling. It has been found that the addition of AEROSIL powder has various beneficial effects on the characteristic of the photopolymer. First of all, the percentage of curing has been found to increase with the percentage of AEROSIL added. This effect can be analyzed with the working curve of the polymer mixture. Moreover, higher viscosity of the polymer mixture reduces the amount of support required when building overhanging parts. Experiments conducted have shown that the mechanical properties of models built with AEROSIL powder mixture have also improved.

4.3.1 Support-less modeling

In the LL process, as parts are fabricated from bottom to top in the liquid photo-polymer, support structures are necessary to build those with floating or over-hanging areas. Data on these supports are included in the part data before the layer-by-layer fabrication, and the supports are removed when the fabrication is completed. The disadvantages of support structures are that: (a) the removal of support structures must be done manually causing accuracy loss; (b) the support structure generation requires some experience, even though there are many software packages which can generate supports for the part automatically; and (c) there exist certain restrictions on the parts that can be fabricated.

One of the ideas to avoid the support structure is to use wax-type photo-polymer as in SGC [29]. This method uses a grinding mechanism to produce a flat surface and it is quite difficult to reduce the size of machine. It also requires the model to be heated to melt down the excess area. The concept of support reduction revolves around the use of a highly viscous material. The viscosity of urethane acrylic-based photopolymer (namely De Solite SCR-310 [30]) can be varied with the addition of fumed silica Degussa AEROSIL powder [31,32]. This powder is used extensively by the paint and epoxy industries for many purposes. One of the main use of the AEROSIL powder is to increase the viscosity of the fluid. The addition of the AEROSIL powder is also associated with the improvement of curing characteristics of the resin, namely the percentage of curing and shrinkage strain after post curing (see ref. [33] for details). It also has the potential to enable the implementation of support-reduction part building through the use of high-viscosity polymer. Resin with higher viscosity will be able to support a thin layer of cured polymer on the surface. Therefore, overhanging parts that require supports in the present method can be built with less support structures, or even without support structures in certain types of model. Hence, materials and preparation time needed for the part building will be greatly reduced. In addition, the element of experience for determining the location of support for overhanging areas can also be minimized for the entire model creation and building process. The reduction in the amount of post-curing shrinkage strain further lessens the amount of support required in the prototyping process.

4.3.2 Layer coating process

Support-less (or minimum support) part building has been attempted using the AEROSIL-mixed photo-polymer [33-36]. Fig. 4.8 shows an experimental setup based on a curtain coating process [37] to extrude the high-viscosity AEROSIL-mixed resin. The mixture of photo-polymer and AEROSIL is extruded from a syringe driven by a motor and released through a slit in the coating head. To ensure homogeneous flow along the slit, a photo-polymer distribution head, consisting of several outlets equally spread across the slit, has been designed and fabricated. The photo-polymer is distributed into the slit of the coating head through these outlets to produce a relatively even curtain coating. With the curtain coating process, higher viscosity of the mixture eliminates the need for settling time in recoating. Hence, fabrication time for a model can be greatly reduced.

4.4 Improvements in LL-process

4.4.1 Mechanical properties

4.4.1.1 Fiber-reinforced LL parts

Molded Specimens

As shown in Fig. 4.9, the stiffness and strength of the fiber-reinforced specimens increase with increase in the fiber concentration in the composite. From the results, it can be seen that specimens with 20% fiber concentrations double the stiffness of their non-reinforced counterpart. The strength of the reinforced specimens is significantly higher than that of the non-reinforced specimen. This is due to the contribution of the short fibers. The increase in the stiffness and strength of the composite is a function of the length as well as the orientation of the fibers.

Fig. 4.9 (a) shows the experimental trend that is similar to the theoretical prediction. However, the theoretically calculated values of the modulus of elasticity are slightly higher than the experimental values for all fiber concentrations tested. This may be due to several factors. The lengths of fiber used have a wide distribution, as shown previously, and not exactly 1.6 mm as specified. The fiber fracture due to handling and during the mixing process and this can also contribute to an increase in much shorter fibers. Fibers with a length shorter than 1.6 mm will contribute to lower reinforcement efficiency, as depicted by Eq.(4.8). A second factor affecting the experimental values may be due to the presence of air bubbles within the structure of the specimen. Air bubbles are generated within the fiber/polymer mixture during the mixing process due to the amount of stirring involved. These air bubbles are trapped within the structure of the composite after curing. The quantity of bubbles was observed to increase with fiber concentration used in the specimens. The entrapped bubbles greatly affect the mechanical properties of the composite due to stress concentration effect from the voids generated.

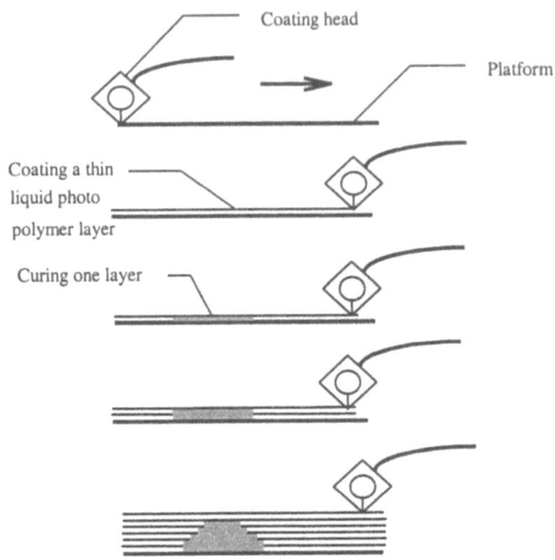

(a) Layer coating concept

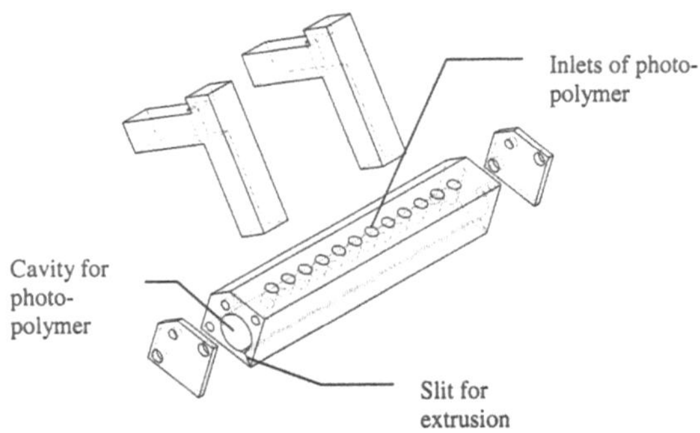

(b) Coating head structure

Fig. 4.8. Layer coating mechanism [36].

LL-built Specimens

The work discussed in this section focuses on specimens with fiber concentrations of 20 % by volume. Although better mechanical properties of the part can be achieved with higher concentration of fibers as reflected by Eq. (4.6) and (4.9), the increase in mixture viscosity caused by the fibers affects the part-building process. These include the material handling process that involves the draining and recycling in LL process. It also affects the layer renewal process as the deposition of flat and thin layers of the mixture for laser scanning is hindered. This will make the fabrication of an accurate and functional LL part impractical.

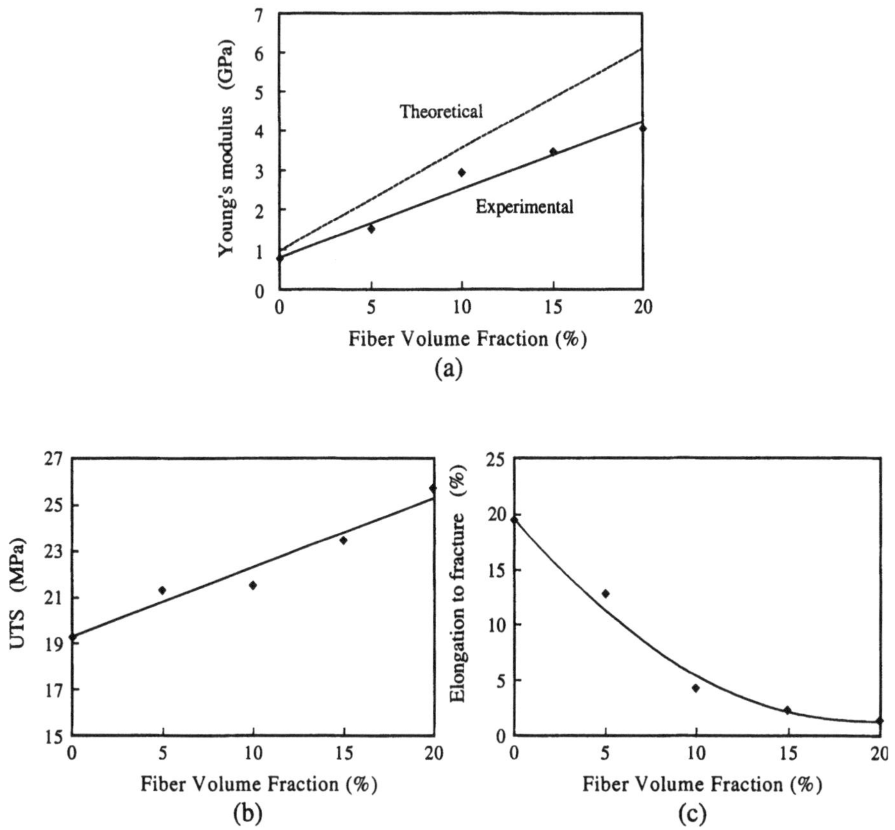

Fig. 4.9 Variation of mechanical properties of the molded SCR310 specimens with fiber concentration.

When fiber concentrations of less than or equal to 20% by volume were used, the accuracy of the built part was found to be within acceptable tolerances. Layer pitch values as low as 150 μm could still be used for part fabrication, hence maintaining high resolution of the built part. However, there is a need to lengthen the waiting time before laser scanning after each layer renewal process to allow the surface to settle.

Fig. 4.10(a) to (c) show the variations of the modulus of elasticity, ultimate tensile strength (UTS) and strain to fracture with laser power for the fiber reinforced SCR310 specimens and the non-reinforced counterparts. The layer pitch was held constant at 150 μm in the fabrication of the specimens. The specimens were post-cured until full curing was attained.

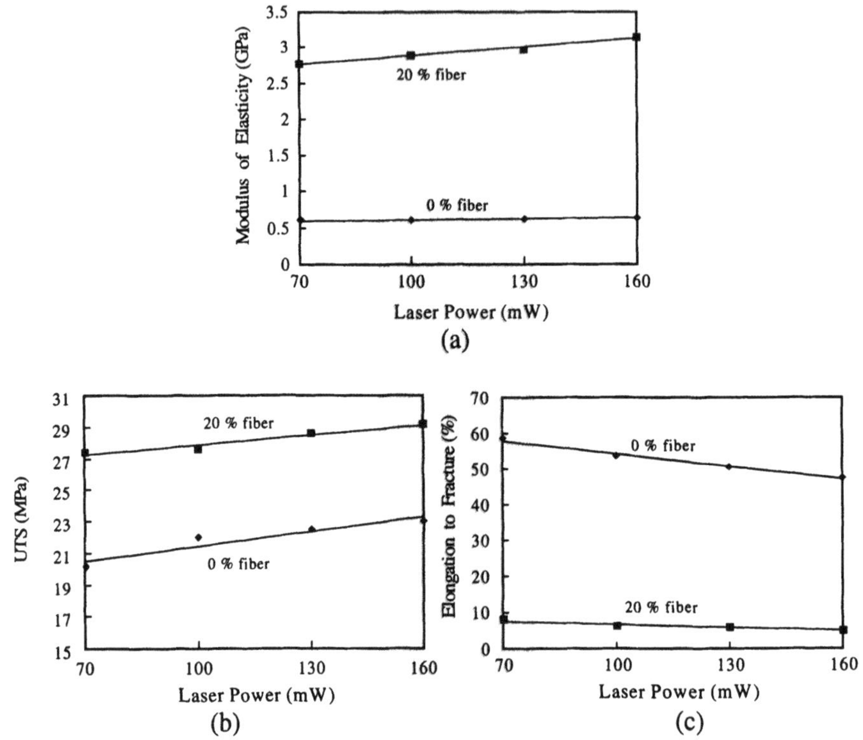

Fig. 4.10 Variation in mechanical properties for LL-fabricated SCR310 specimens with 20% and 0% fiber concentrations, at a layer pitch of 150μm.

The upward trend shown in Fig. 4.10 (a) and (b) indicates that an increase in the strength and stiffness of the fiber-reinforced specimens can be introduced by using higher laser powers and smaller layer pitches during fabrication. This is due to the increased amount of initial curing brought about by the use of such parameters during fabrication. The use of these parameters will also bring about a more compactly cured structure, leaving little space for partially cured and uncured polymer to be retained, and hence lowering the amount of post-curing induced shrinkage. However, the specimens become more brittle when such parameters are used during the fabrication process, as observed in Fig. 4.10 (c).

4.4.1.2 AEROSIL-mixed LL parts

The UTS of the specimen increases with the AEROSIL percentage at both the green and post-cure states. A comparison of the green-state and post-cured specimens is shown in Fig. 4.11. The UTS for a sample with the same amount of AEROSIL powder mixing is higher after post-curing in UV chamber. This result is expected, as the post-curing process will polymerize the uncured or under-cured photo polymer in the built part. With a more thoroughly cured structure, the UTS is expected to be higher.

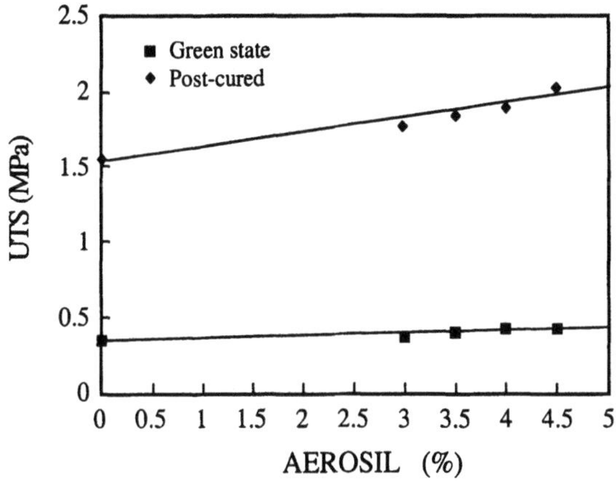

Fig. 4.11 UTS of model with different percentages of AEROSIL mixture (scan speed of 983 mm/s; the layer pitch of 200 μm; and the scan pitch of 200 μm).

These observations are in accordance with the results showing the increase in cure depth and line-width at a higher percentage of AEROSIL powder [35]. The increase in UTS is due to a larger C_d and L_w during the photo fabrication process. E_c and D_p, however, decrease with the percentage of AEROSIL added, as shown in Table 4.1.

Table 4.1 Critical exposure and penetration depth.

Percentage of AEROSIL (%)	Critical Exposure E_c (mJ/cm^2)	Penetration Depth D_p (μm)
3.00	11.785	161.68
3.50	10.336	155.64
4.00	8.844	148.75
4.50	6.911	139.42
5.00	5.926	136.30

The addition of AEROSIL can improve both the C_d and L_w. These imply that at a higher percentage of AEROSIL powder mixing, there will be more overlapping of the cured line in the structure of the built parts. Hence, less uncured and under-cured photo polymer remain within the parts. The overlapping of the cured lines is the fundamental process which contributes to the strength and integrity in LL parts. Therefore, the addition of AEROSIL powder to the photo polymer increases the overall strength of the completed part as it enhances the overlapping process in the photo fabrication.

4.4.2 Shrinkage and distortion

Fiber-reinforced

Fig. 4.12 shows the longitudinal and transverse shrinkages for both the fiber-reinforced SCR310 specimens and the non-reinforced counterparts at different laser powers and layer pitches used during fabrication. A common trend observed from both graphs is that the longitudinal and transverse shrinkages are functions of the laser power and layer pitch. The shrinkages in both directions decrease with the use of high laser power and low layer pitch settings. By comparing the magnitudes of the shrinkage for both graphs, the fiber-reinforced specimens are seen to encounter less shrinkage compared to the non-reinforced specimens when fabricated using low layer pitches and high laser powers. The reduced shrinkage may be due to the fact that the fibers added into the specimen displace some of the volume originally occupied by the partially cured and uncured resin within the structure after laser fabrication. The smaller amount of partially cured and uncured resin results in a reduction in the shrinkage encountered by the fiber-reinforced specimens upon post-curing.

However, fiber-reinforced specimens built with a high layer pitch and a low laser power exhibit higher amount of shrinkage compared to the non-reinforced counterparts. The fibers within the resin cause the resin mixture to become more opaque to laser penetration. Refraction and reflection of the laser beam occur at the fiber/resin interfaces, reducing the penetration depth of the laser beam. Using a low laser power during fabrication will result in the layers below each layer to be poorly cured due to weaker penetration. The larger amount of partially cured polymer in the lower portion of each fabricated layer will bring about increased shrinkage encountered by the part during the post-curing process, resulting with an end product of poor dimensional accuracy.

Unlike the trend exhibited by the pure resin specimens in Fig 4.12(b), at low laser powers, longitudinal shrinkage of the fiber-reinforced specimens is found to exceed that in the transverse direction (Fig. 4.12(a)). By using a higher laser power, longitudinal shrinkage is much lower than transverse shrinkage. The increase in the longitudinal shrinkage, when using a lower laser power, is caused by the further polymerization of the higher amounts of liquid and partially cured polymer retained in the lower portion of each layer in the fiber-reinforced specimens.

Comparing the above observation with that for the non-reinforced specimens, a different trend in the shrinkage can be observed. Unlike the reinforced specimens, where directional domination is dependent on the parameters used, the transverse shrinkage is seen to pre-dominate over longitudinal shrinkage for the whole range of parameters tested. However, it should be noted that if the laser power was to be increased above that used for the current experiments on the non-reinforced specimens, longitudinal shrinkage might have pre-dominated over transverse shrinkage as exhibited by the SCR300 non-reinforced specimens discussed in section 3.1.4.

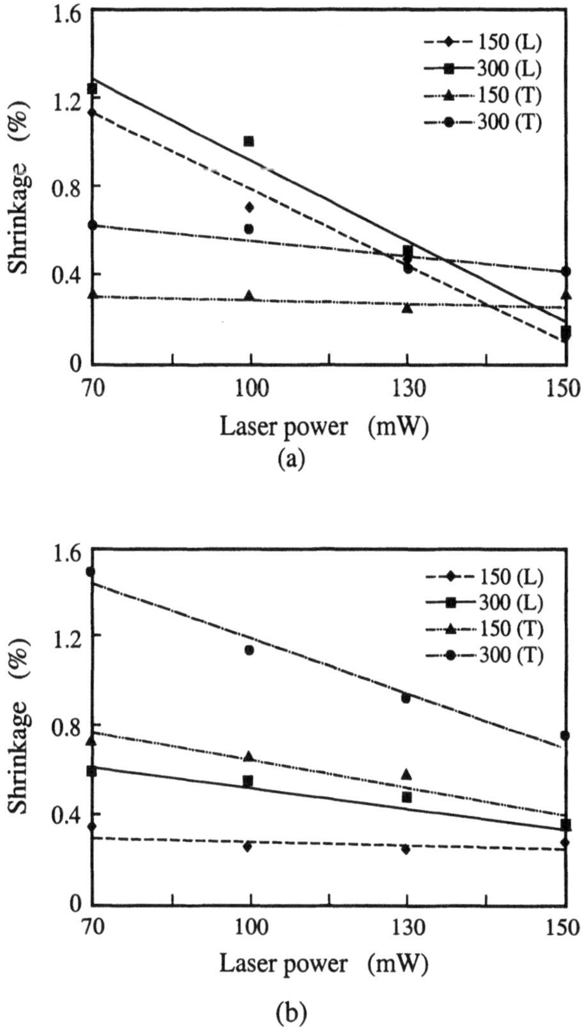

Fig. 4.12 Longitudinal (L) and transverse (T) shrinkages for the (a) fiber-reinforced specimens and their (b) non-reinforced (pure resin) counterparts at different laser powers and layer pitches

From Fig. 4.12(a), the steeper trend in the graph observed for longitudinal shrinkage compared to transverse shrinkage indicates a higher influence brought about by the change in process parameters. At low layer pitch and low laser power settings, larger amount of partially cured resin is contained within the specimens. These parameter settings contribute to the high values of longitudinal shrinkage encountered during post curing. Furthermore, due to the existence of fibers within the specimens, most of the cavities caused by the "Gaussian" profile of the laser beam are eliminated by displacement effect, as evident from Fig. 4.5. This reduces the uncured liquid resin entrapped within the specimen, which is largely responsible for the transverse shrinkage encountered by the specimens during post-curing. Because of this, the transverse shrinkage is not greatly affected by the change in the laser power used, as evident from the graph plotted. The above explains the difference in trends for the type of shrinkage that is seen to predominate at different parameter settings for both the fiber-reinforced and non-reinforced specimens.

AEROSIL-mixed

Fig. 4.13 illustrates the longitudinal and transverse shrinkage strains of the specimens with dimensions of 20mm x 20mm x 5mm for different percentages of AEROSIL. Both the longitudinal and transverse shrinkage strains decrease with the percentage of AEROSIL. This may be due to the added AEROSIL powder displacing some of the volume originally occupied by the partially cured and uncured polymer within the part. The reduced amount of the partially cured and uncured polymer results in a reduction in the shrinkage strain upon post curing.

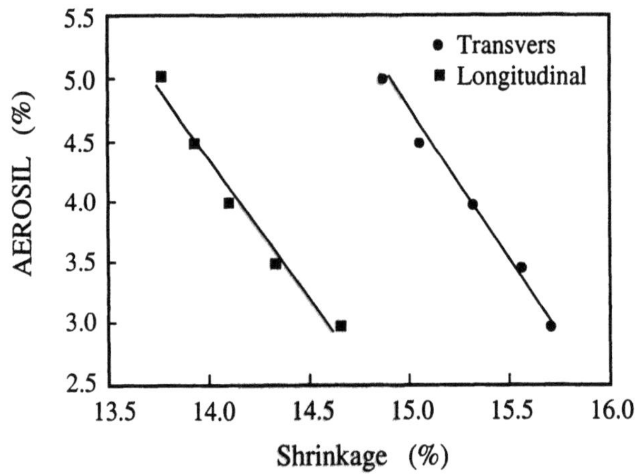

Fig. 4.13 Percentage of AEROSIL vs. percentage of shrinkage.

4.4.3 Reduction of support and build time

Increased Degree of Curing

In Chapter 2, it has been shown that the cured line-width of the photo polymer is a function of C_d and D_p. From reported experiments [35], it has been shown that the AEROSIL-mixed photo-polymer also obeys the Beer-Lambert Law of absorption. As a result, the last column of Table 4.2 shows that the C_d / D_p ratio increases as the percentage of AEROSIL powder increases.

Table 4.2 Critical depth and penetration depth with the percentage of AEROSIL added.

Percentage of AEROSIL	Cure Depth C_d (μm)	Penetration Depth D_p (μm)	Ratio of $C_d/2D_p$
3.00	374.0	161.68	1.1566
3.50	380.5	155.64	1.2224
4.00	386.0	148.75	1.2975
4.50	396.0	139.42	1.4202
5.00	410.0	136.30	1.5040

Recall from Chapter 2, the line-width function is:

$$L_w = \sqrt{2}\, W_0 \left[ln \left({}^{E_{max}}\!/_{E_c} \right)^{1/2} \right] \tag{4.11}$$

and also from

$$C_d / D_p = ln\left(E_{max}/E_c\right) \tag{4.12}$$

Substituting Eq. (4.12) into Eq. (4.11), we obtain:

$$L_w = 2W_0 \sqrt{C_d / 2D_p} \tag{4.13}$$

The cured line-width L_w is directly proportional to the laser spot diameter $2W_0$. The cured line-width is also proportional to the square root of the ratio of the cured depth to the photo polymer penetration depth (C_d/D_p). When the ratio of C_d/D_p is substituted into Eq. (4.13), the resulting L_w will be a function of the percentage of AEROSIL powder in the photo-polymer. It is found that for the same curing conditions, L_w increases with the percentage of mixing. This results in wider cured lines along the scanning direction during photo-polymerization. The wider cured line overlaps more with the previously cured line, leading to a higher percentage of curing in the overlapping region. Hence, a more compact cured structure with less retained uncured resin is fabricated.

The degree of curing at different percentage of mixing is shown in Fig. 4.14. It clearly illustrates that as the percentage of added AEROSIL increases, the percentage of curing in the specimen increases. The reason for the increase in percentage of curing is due the to the increase in cured line-width L_w of the photo polymer.

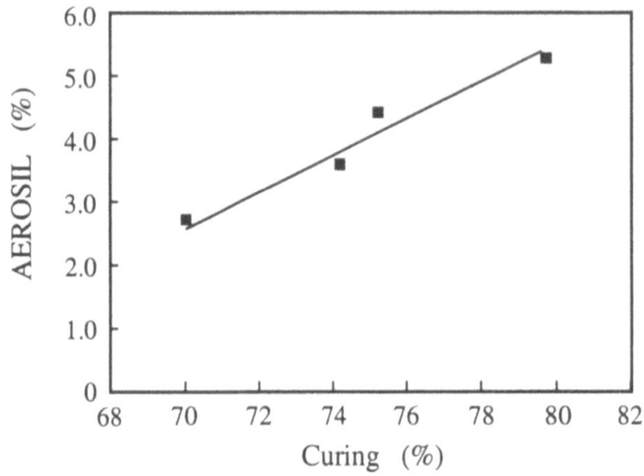

Fig. 4.14 Percentage of AEROSIL vs percentage of curing.

Less Support and Build Time

As mentioned earlier, the percentage of curing has been found to increase with the percentage of AEROSIL added. With a curtain coating process, the higher viscosity of the mixture significantly reduces the settling time in recoating. Hence, the fabrication time for a model can be greatly reduced. Moreover, the higher viscosity of the polymer mixture reduces the amount of support required when building overhanging parts.

Fig. 4.15 shows models built with the new curtain coating process. From these built models, three distinctive improvements due to the curtain coating process have been identified: the model building speed, the physical model and support-less modeling. The test model (a cup with an inner diameter of 34.5mm and a height of 35.7mm) exhibits less distortion. The overhang area

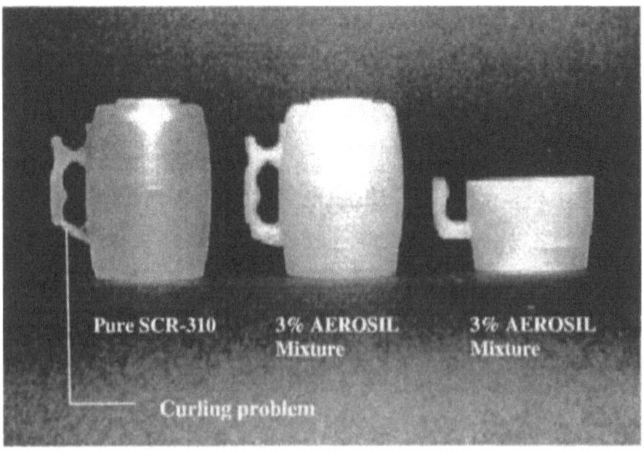

Fig. 4.15 Model built without supports from pure photo-polymer and AEROSIL-mixture, respectively.

was also built without the supports while maintaining a comparable quality as that of a model built with support. Table 4.3 compares the building time with pure polymer and with AEROSIL mixture. The increase in the model building speed for the curtain coating process is achieved by the elimination of the settling time.

Table 4.3. Building time comparison [34].

Time Taken Per Layer	Pure SCR-310	AEROSIL Mixture
Platform Motions (sec)	5.00	0.50
Down Time (sec)	5	0
Settling Time (sec)	30	1
Scanning Time (sec)	15.23	15.23
Coating Time (sec)	2	8
Layer Renewal (sec)	57.23	24.73
Total Building Time (min)	113	49

4.5 Summary

Part shrinkage and weak mechanical properties are some of the major problems encountered in rapid prototypes fabricated using photo-polymers in the LL process. The mechanical properties of the LL-built parts can be improved by the introduction of short glass fibers to the polymer material used. From experimental results, it was found that the specimens fabricated with the inclusion of glass fibers showed better strength compared to those fabricated using other powders-mixed.

The improved LL method using the AEROSIL–mixed polymer can reduce the need for support structures since its high viscosity support the over-hangings. Many other good features of this method include higher tensile strength, lower shrinkage, and shorter fabrication time. Although the higher percentage of the AEROSIL can give better features, it requires high pressure to be extruded from the thin slit in the dispensing head for the curtain coating process, and air bubble trapping problem may become more evident.

4.6 References

1. C.M. Cheah, J.Y.H. Fuh, A.Y.C. Nee, et al., *J. Mater. Proc. Techn.*, Vol. 67 (1997), 41.
2. C. Decker and K. Moussa, *J. Coatings Techn.*, Vol. 65(819), (1993), 49.
3. C. Decker and K. Moussa, *J. Polym. Sci., Polymer Lett. Ed.*, Vol. 27 (1989), 347.
4. P. Karrer, S. Corbel, J.C. Andre, and D.J. Lougnot, *J. Polym. Sci., Polymer Chem.*, Vol. 30 (13), (1992), 2715.
5. L. Lu, J.Y.H. Fuh, A.Y.C. Nee, E.T. Kang, T. Miyazawa, and C.M. Cheah, *Mater. Res. Bulle.*, Vol. 30(12), (1995), 1561.
6. G. Bugeda, M. Gervera, *RP. J.*, MDB University Press, University of Nottingham, Nottingham, UK, Vol. 1(2), (1995), 13.
7. B. Wiedemann, K.H. Dusel, and J. Eschel, *Rapid Prototyping J.*, MDB University Press, University of Nottingham, Nottingham, UK, Vol. 1 (3), (1995), 17.

8. C.C. Chen, and P.A. Sullivan, *Proc. of the 6th Inter. Conf. on Rapid Prototyping*, Dayton, OH, (1995), 153.
9. S. Das, N. Harlan, J.J. Beaman, D.L. Bourell, *Proc. of Solid Freeform Fabrication Symposium*, UOT, Austin, Texas, (1996), 89.
10. M. Wohlert, D. Bourell, *Proc. Solid Freeform Fabrication Symposium*, UOT, Austin, Texas, (1996), 79.
11. U. Hejmadi, K. McAlea, *Proc. Solid Freeform Fabrication Symposium*, UOT, Austin, Texas, (1996), 97.
12. C. Greer, J. McLaurin, A.A. Ogale, *Proc. Solid Freeform Fabrication Symposium*, UOT, Austin, Texas, (1996), 307.
13. P.F. Jacobs, *RP. and Manuf.: Fundamentals of Stereolithography*, 1st ed., Society of Manufacturing Engineers, Dearborn, Michigan, (1992).
14. C.M. Cheah, *PhD. Dissertation*, national University of Singapore, (2001).
15. D. Hull, *An Intr. to Comp. Mater.*, Cambridge University Press, ed. 2 (1981).
16. J.C. Halpin, J.L. Kardos, *Polym. Eng. Sci.*, Vol. 18 (6), (1978), 496.
17. J.E. Ashton, J.C. Halpin, P.H. Petit, *Primer on Composite Analysis*, Technomic Publishing Co., Ch. 5 (1969).
18. J.C. Halpin, J.L. Kardos, *Polym. Eng. Sci.*, Vol. 16, (1976), 34.
19. P.E. Chen, *Polym. Eng. Sci.*, Vol. 11, (1971), 51.
20. R.M. Barker, T.F. MacLaughlin, *J. Comp. Mater.*, Vol. 5 (1971), 49.
21. V.R. Riley, *J. Comp. Mater.*, Vol. 2, (1968), 436.
22. R.F. Gibson, *Principles of Comp. Mater. Mech.*, McGraw-Hill, (1994).
23. H.L. Cox, *British J. of Appl. Phys.*, Vol. 3, (1952), 72.
24. A. Kelly, A., W.R. Tyson, *J. Mech. Phys. of Solids*, Vol. 13, (1965), 329.
25. H. Krenchel, *Fiber Reinforcement*, Copenhagen, Akademisk Forlag (1964).
26. A. Kelly, W.R. Tyson, *High Strength Material*, (1965), 578.
27. D.K. Hale, A. Kelly, *Annual Rev. Mater. Sci.*, Vol. 2, (1972), 405.
28. C. M. Cheah, *RP. J.*, Vol.5 (3), (1999), 112.
29. *Solid Ground Curing*, Cubital, Ltd., Raansna, Israel (2001), www.cubital.com.
30. *SCR-310*, Brand Name of UV Curable Resin from Nippon Synthetic Rubber Ltd., Japan, (1995).
31. *Technical brochure for AEROSIL*, Fumed Silica, Degussa, Germany,1997.
32. C.S. Wen, *The Fundamentals of AEROSIL Dynamics*, World Scientific Publishing Co. Inc., New Jersey, (1996).
33. Y.S. Lim, Y.S. Wong J.Y.H. Fuh, T. Miyazaw, *Proc. of, the 8th European Confer. on RP. and Manuf.*, (1999), 313-24.
34. J.Y.H. Fuh, Y.S. Wong, Y.S. Lim, T. Miyazawa, A.Y.C. Nee, and H.T. Loh, *Proc of The 8th Inter. Confer. on RP.*, (2000), 123.
35. Y.S. Lim, *MEng Thesis*, National University of Singapore, (2001).
36. T. Miyazawa, *PhD Dissertation*, National University of Singapore (2001).
37. J.P. Kurth, I. Meyvaert, P. Vandormae, *Proc. of the 7th Inter. Conf. On Rapid Prototyping*, San Francisco, (1997), 218.

Chapter 5

SELECTIVE LASER SINTERING

5.1 Principle of laser sintering

5.1.1 Fundamentals of laser processing

For a given material system, the laser energy density, E_s is a key factor which affects quality of the laser sintered structure. A laser energy density is defined by the laser power, laser scan speed and laser beam spot size:

$$E_S = \frac{P}{v\delta} \qquad (\text{J/cm}^2) \qquad (5.1)$$

where P is the incident laser power (Watts), v, the laser scan speed (cm/s) and δ, the laser beam spot diameter (cm).

From Eq. (5.1), it is understood that the laser energy density is not only dependent on the laser power, but also on the laser scan speed and beam spot size. Therefore, laser energy density, rather than laser power, should be considered in laser processing.

5.1.2 Absorptance of laser energy

In the laser application, the process depends on the laser energy density as well as the thermo-physical parameter, which is the normal spectral absorptance of the laser beam. The absorptance of a material, A, is defined as the ratio of the absorbed radiation to the incident radiation [1]:

$$A = 1 - R$$

where R is the reflectance that is the ratio of the reflected radiation to the incident radiation.

Since the entire thermal cycle occurs locally and very quickly, typically between 0.5 and 25 ms, a successful process using a laser technique depends strongly on adequate absorption or coupling of the laser energy [2]. Table 4.1 lists the measured absorptance using Nd-YAG and CO_2 lasers for different elements and compounds.

Table 4.1 Absorptance of elemental and compound powders, Nd-YAG ($\lambda = 1.06\ \mu m$) and CO_2 ($\lambda = 10.6\ \mu m$) [1]

Material	Absorptance $\lambda = 1.06\ \mu m$	Absorptance $\lambda = 10.6\ \mu m$
Cu	0.59	0.26
Fe	0.64	0.45
Sn	0.66	0.23
Ti	0.77	0.59
Pb	0.79	
Co-alloy (1%C; 28%Cr; 4%W)	0.58	0.25
Cu-alloy (10%Al)	0.63	0.32
Ni-alloy I (13%Cr; 3%B; 4%Si; 0.6%C)	0.64	0.42
Ni-alloy II (15%Cr; 3.1%Si; 0.8%C)	0.72	0.51
ZnO	0.02	0.94
Al_2O_3	0.03	0.96
SiO_2	0.04	0.96
BaO	0.04	0.92
SnO	0.05	0.95
CuO	0.11	0.76
SiC	0.78	0.66
Cr_3C_2	0.81	0.70
TiC	0.82	0.46
WC	0.82	0.48
$NaNO_3$	0.16	0.80
NaCl	0.17	0.60
Polytetrafluoroethylene	0.05	0.73
Polymethylacrylate	0.06	0.75
Epoxypolyether-based polymer	0.09	0.94

The reflectivity of a material is most dependent on its electrical conductivity. A metal with high electrical conductivity has the highest reflectivity; for example copper. Hence, a high energy density is required to sinter a material like copper. A material with high thermal diffusivity will normally allow a greater depth of fusion penetration with no thermal shock or cracking. In general, materials in powder form exhibit higher absorptance than their bulk form [1]. In the case where a mixture of powders is used, the absorptance of the mixture A_{mix} can be expressed by the rule of mixture [1]:

$$A_{mix} = A_1 f_1 + A_2 f_2 \tag{5.2}$$

where A_i and f_i are the absorptance and volume fraction of component i, respectively.

In addition, the thermo-physical property of the material is affected by the laser process [1]. As shown in Fig. 5.1, the absorptance of Ni rises quickly due to changes in the thermo-physical properties of the powder at the beginning of the laser heating (zone I). If laser heating stops immediately, the absorptance returns to its initial magnitude after cooling. After prolonged heating, the absorptance value has been observed to increase further due to melting and particles rearrangement. With longer heating, the powder porosity decreases, slowing down the increase in the absorptance (zone II). However, the SLS process is fast so that the absorptance can be considered as constant within a single scanned track. For multi-scanned tracks and layers, the rise in temperature will affect absorptance.

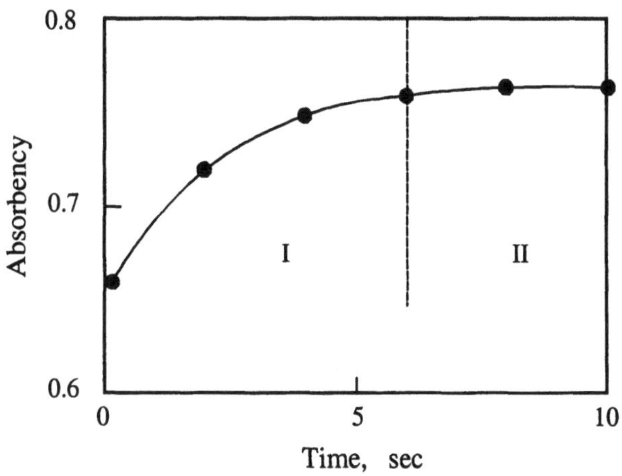

Fig. 5.1 Variation of absorptance during laser process of Ni alloy [1].

5.2 Process of selective laser sintering

Parts fabricated by the SLA technology is in general unable to be directly used as functional parts due to the nature of the polymers. To overcome this, the University of Texas at Austin has developed a technique called selective laser sintering (SLS) [3-5]. SLS is also a rapid prototyping process, similar to SLA that creates three-dimensional freedom objects directly from their CAD models. However, the part is created by selectively fusing thin layers of metal powder with a scanning laser beam. This process is later commercialized by DTM Corp of Austin, Texas. EOS GmbH at Munich, Germany also commercialized a so-called direct metal laser sintering (DMLS). Its main application is the manufacture of mold inserts for plastic injection molding [6].

5.2.1 Process

SLS-built solid parts were initially used primarily to enhance communication as well as allow testing of concepts during the design cycle. Currently SLS is used to produce parts for tooling and mold production. In many cases SLS has a distinct manufacturing speed and cost advantage over conventional methods for producing prototypes and short-run tooling. The greatest potential of the SLS process may lie in the manufacturing of specialized and functional products.

Fig. 5.2 schematically shows the principle of SLS process. This process begins with the deposition of a thin layer of powder onto the surface of a substrate. The powder layer is then leveled by a roller rolling over the powder bed. After rolling, the powder surface is raster-scanned with a high power energy laser beam. The beam intensity is modulated to fuse the powder in areas to be occupied by the part at that particular cross-section. Sintering or bonding between particles during laser raster-scanning occurs due to high temperature generated by the laser energy, which leads to melting, partial melting or softening of particles, or of coating on the surface of the particles. Successive layers of powder are then deposited and sintered until the entire part is produced. Each layer is sintered deeply enough to fuse it to the underlying layer. The next layer of powder material is then spread out, leveled and selectively sintered over the top surface of the growing structure.

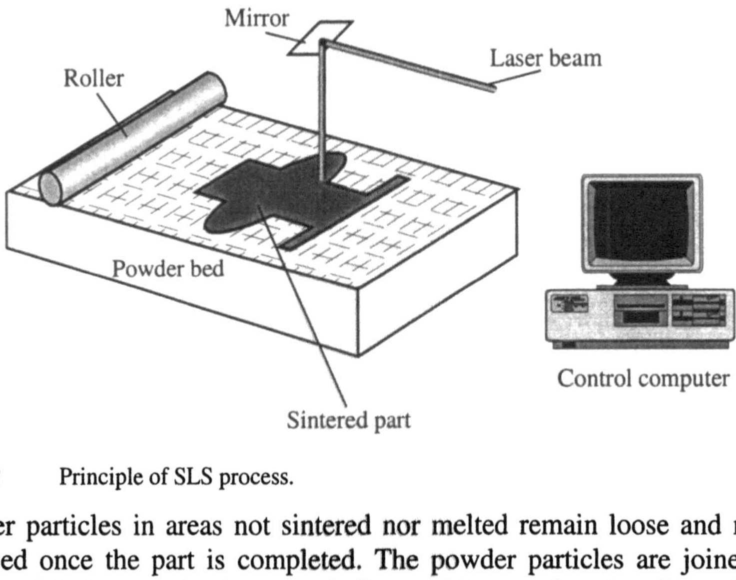

Fig. 5.2 Principle of SLS process.

Powder particles in areas not sintered nor melted remain loose and may be removed once the part is completed. The powder particles are joined by a variety of fusion mechanisms, including melting, surface bonding, sintering aids, and polymer coatings. The unfused material remains in place as the support structure that is important for three-dimensional building. When the shape is completely built up, the part is separated from the loose supporting powder.

Several types of material systems are commercially available, including plastics, waxes, and low melting metal alloys, as well as polymer coated metals and ceramics for making green parts. Prototype parts have been produced using various thermoplastic, metal composite and ceramic composite powders [7-9].

5.2.2 Types of SLS

Selective laser sintering process can be further divided into two different processes, namely indirect and direct selective laser sintering.

5.2.2.1 Indirect SLS

In indirect SLS, DTM has developed metallic and ceramic powders coated with a 5 μm thick layer of polymer. The polymer layer is only melted during the SLS process to form a green part. The powder particles are actually bonded together by the polymer, as shown in Fig. 5.3.

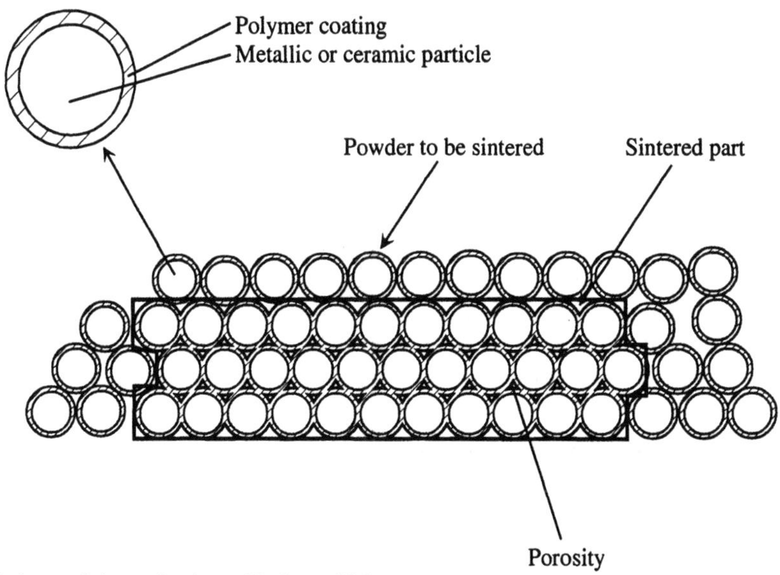

Fig. 5. 3 Schematic view of indirect SLS.

Since only the polymer layer is melted, the green part fabricated by indirect SLS has high porosity, as shown in Fig. 5.3, and cannot be used without further post process. Basically, the green parts have to be infiltrated with a water soluble polymer to enable a stepwise debinding of the green part. In the final step, the part is debinded and sintered in combination with the infiltration of Cu. The debinding process causes shrinkage of 4.5% horizontally and 5% vertically[10].

5.2.2.2 Direct SLS

In direct SLS, a high-energy laser beam directly consolidates the metal powder to a high density (>80% theoretical density), preferably with minimal or no post-processing requirements. The main advantages associated with this technique are the elimination of expensive and time-consuming pre-processing and post-processing steps, as compared to indirect or conventional processes. Direct SLS has been investigated on metal and metal-ceramic material systems in the past [3]. These material systems include binary and ternary mixtures, such as bronze-nickel and tungsten carbide-cobalt-nickel. Density of up to 80% of the theoretical density can be attained. Post-processing of direct SLS bronze-nickel parts by conventional liquid-phase sintering at elevated temperatures to increase the density to 95% of the theoretical density has been reported. Post-processing of SLS bronze-nickel parts by hot isostatic pressing has been shown to attain nearly full density (>98%). The material systems investigated in the past were mainly liquid-phase sintering systems composed of at least two phases.

Fig. 5.4 shows different alloying techniques used in SLS [11]. Fig. 5.4 (a) shows matrix powder particles fully coated with a different alloying material. Sometime a mixture of different powders is used. It is easy to adjust the alloy composition by varying the ratio between different elemental or alloy powders (Fig. 5.4 (b)). However, it is important to ensure a homogenized powder mixture due to the differences in density and size of the powders. If the powder particles are not well mixed, segregation of different elements can happen after melting. Separation of the different powders can also take place during the deposition of the powder onto the powder bed. Partially coated powders can be achieved by different means, such as by mechanical milling of two or more elemental or alloyed powders together (Fig. 5.4 (c)). Segregation of the partially coated powder can be avoided since the different powder particles have already been mechanically bonded together before SLS, as shown in Fig. 5.4 (d). If the powders are mechanically milled for too long time, the alloying elements may be fully embedded into the matrix powder. It may not be advantageous if the alloying element has low melting point.

The advantage of using a mixture of different components is to eliminate post processing. The laser only melts the compounds with low melting temperature. The molten compounds flow with the help of capillary forces through the pores among the solid particles. For better densification, the particle size of the low melting powder should be smaller than that of the structural materials because the bigger particle has high melting enthalpy due to the larger mass [12].

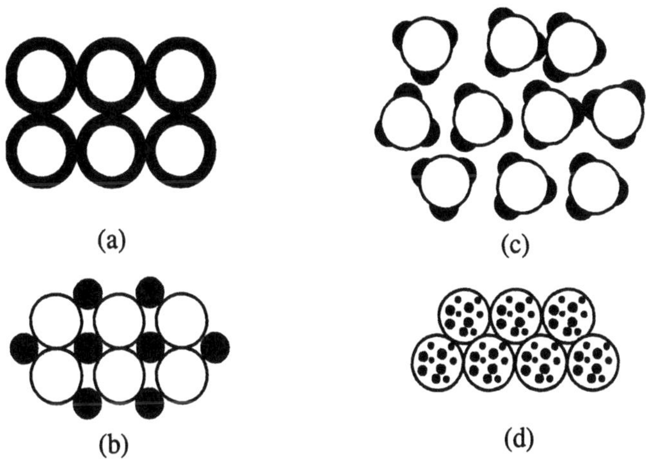

Fig. 5.4 Different kinds of powders for SLS process: (a) coated powder, (b) mixture of powders, (c) partially coated powder and (d) fully mixed powder [11].

Table 5.2 lists some combinations of metals used in the two-component process [12]. Basically, the systems consist of two elements. One of the two in the system has a low melting point. The possible resultant compounds are also given in the table.

Direct SLS is also possible with high performance materials such as nickel and cobalt base superalloys, cermets, titanium base alloys and monolithic high temperature metals such as molybdenum. These materials are used for high performance components that typically experience high operating temperatures, high stress, and severe oxidizing or corrosive environments [13].

Table 5.2 Combinations of metals and possible compounds that can be formed.

Element	Melting point (°C)	Possible compound
Ni/Sn	1455/232	$Ni_3Sn/Ni_3Sn_2/Ni_3Sn_4$
Fe/Sn	1540/232	$FeSn/FeSn_2$
Cu/Sn	1083/232	$Cu_{31}Sn_8$, Cu_6Sn_5, Cu_3Sn
Fe/Cu	1540/1083	Fe_4Cu_3
Ni/Cu	1455/1083	α_1 and α_2 solid solutions

Commercial direct SLS equipment is currently available, which uses nickel-bronze-copper phosphide (bronze-Ni-Cu_3P) powder without using polymeric binder [14]. With this process, parts or molds can be practically fabricated without any post-sintering.

Important parameters influencing the performance of these equipment are the material, the optical beam guidance system, the powder delivery system, and the laser irradiation strategies. Post-processing operations such as post-sintering, infiltration, and bead blasting, are also important.

Though any powdered metallic material may theoretically be used for SLS [15], part shrinkage, porosity, thermal and mechanical stability are very important issues. For instance, bronze-Ni powder exhibits a near-zero shrinkage with minimal curl and warpage [6]. Metallic molds for injection molding have been successfully fabricated with commercial equipment using the bronze-Ni powder [6, 14] or the polymer-coated metallic powder [16]. However, the limited strength, wear resistance and surface quality of such parts make it necessary to employ post-processing, such as binder burn-out and post-sintering (in the case of polymer coated powders), infiltration with resin or low melting point alloys [17], and polishing. The use of such molds is generally limited to the manufacture of a few hundred to thousand parts [6, 14].

By removing the polymer from the powder, direct SLS may provide short processing time and high density parts. SLS appears to be a promising method. Few materials are effectively available, and post-processing is still often unavoidable.

The direct SLS process melts the component matrix to obtain the appropriate amount of flow from the molten material. The liquid-phase particle penetrates the surrounding cavities and wets the bronze/nickel particles [18]. An appropriate amount of flow is critical and can be described as the flow that eliminates porosity, producing a highly dense part and maintaining dimensional accuracies. The appropriate amount of flow is controlled by factors such as atmosphere, powder bed temperature and laser energy density. The three parameters in energy density are the laser power, scan line spacing (scan pitch) and scan speed [19].

If the energy density is too high the surface begins to vaporize before a significant depth of molten material is produced. The sintered layer thickness decreases with increasing scan speed due to the shorter interaction (sintering) time [3]. This thickness also decreases with decreasing scan pitch if the laser beam spot is larger than the pitch. The smaller the scan pitch, the more scan overlapping will occur. The thermal conductivity and reflectivity of the sintered solid are higher than those of the starting powder. The more the scan overlapping occurs, the more the laser energy will be transferred away by heat conduction through the sintered solid and reflected away by the sintered solid surface to result in a decrease in the layer thickness.

Basically larger particles have less tendency to melt than the smaller ones. For example, in the bronze-Ni system, bronze may partially melt at a higher temperature. The bronze and Ni homogenize at high temperatures, either by inter diffusion if the homogenization is between solid bronze and solid Ni, or by solution reprecipitation if the homogenization is between liquid bronze and solid nickel, to form a homogeneous solid solution of Cu in Ni [2]. Under a certain condition, the growth or swelling of the sintered part may occur [20].

By using the right parameters, the typical shrinkage due to the liquid-phase sintering can be completely compensated by an expansion of the material caused by the diffusion of the component, so that the material has no net volume change during the laser sintering process [21]. Due to the small laser focus of 350μm and the high laser scanning speed of 300-800 mm/s, the duration of heating is short enough to avoid oxidation even in a non-inert gas environment. In addition, the phosphor content inhibits oxidation of the powder mixture and also improves the wetting of the solid particles in the molten phase. The wetting and the diffusion are very important mechanisms to improve the mechanical properties of the laser-sintered parts. It should also be noted that the most important variations are an increased content of liquid phase, an enlarged surface area for the diffusion, and an optimization of the laser parameters.

5.2.2.3 Deoxidization

When the volume percentage of melt increases, atmospheric control becomes important in minimizing oxygen pick up in the liquid and oxide formed on the liquid and solid surfaces. Oxygen pick-up lowers the liquid surface energy. The metal oxides have much lower surface energy than the associated metal, leading to a decrease in wettability. The formation of the oxide layers also induces impurity and defects in the SLS-processed parts. Therefore, for the direct SLS process, the atmosphere must be controlled. Depending on the material systems, nitrogen, argon, nitrogen-hydrogen mixture and hydrogen have been used as protective/reducing atmospheres. By using different protection gases, the wetting can also be enhanced.

Additionally, various types of deoxidizer coating or components are used to enhance the wettability of SLS material systems. The principle of deoxidizer is to reduce oxide layer on the surface of the powder particles. These deoxidizers include $ZnCl_2$ [11], $NaCO_3$, $KCrO_3$ [22], Si, and H_3BO_3 [23],

The influence of the deoxidizers on the contact angle between liquid Cu and solid Fe is shown in Fig. 5.5. It demonstrates that the increase in Si and H_3BO_3 can increase wetting between Cu and Fe. Si and H_3BO_3 absorb oxygen by forming SiO_2 and B_2O_3 since SiO_2 and B_2O_3 have much negative Gibbs free energy of -130 and -135 kg.cal/g.ml respectively, compared to -72, -56, -48, -24 and 3 respectively for FeO, Fe_2O_3, SnO, Cu_2O and CuO [23]. Yang *et al.* [23] also found that the addition of Si and H_3BO_3 can reduce the number of gas holes.

5.2.2.4 Effect of laser sintering parameters

The laser power and laser scan speed strongly affect the properties of sintered parts. Fig. 5.6 shows the influence of the laser powder and scan speed on the density of the bronze-Ni parts [2]. The density of the SLS processed bronze-Ni

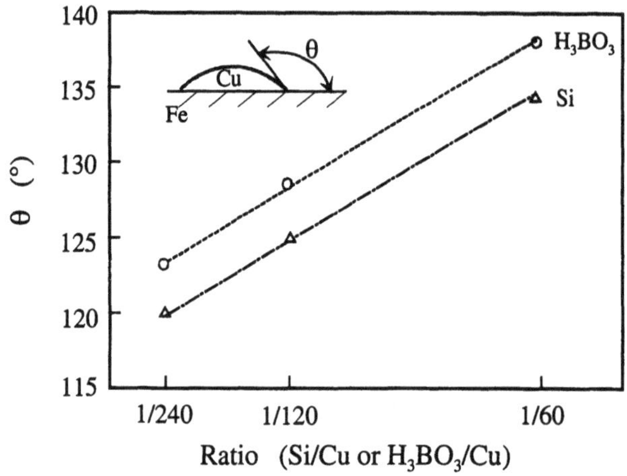

Fig. 5.5 Influence of addition of deoxidizer Si and H_3BO_3 on the contact angle between liquid Cu and solid Fe [23].

parts increase with decrease in the scan speed as well as the laser power. A higher laser density can be achieved at slower scan speed and higher laser power due to an increased amount of energy input to the powder surface. Hence, a large amount of liquid phase is formed, resulting in higher densification.

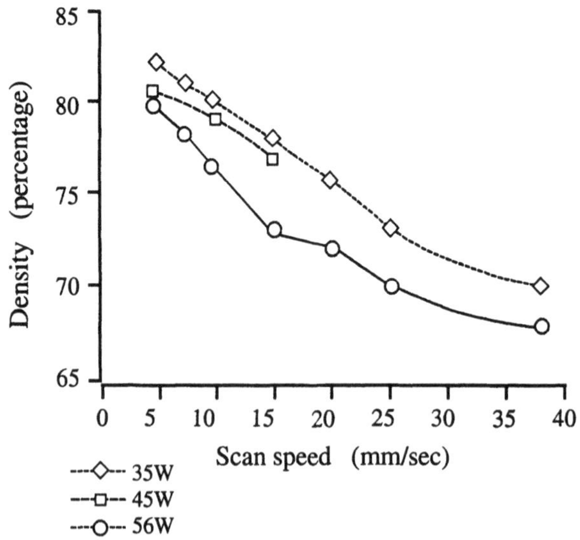

Fig. 5.6 Effect of laser power and scan speed on the density of bronze-Ni parts [2].

The density of SLS processed parts can also be increased by using a mixture of powders with different sizes because the smaller particles can fit into the interstices between the larger particles. The density of a SLS processed part is also affected by its initial packing density. In a high starting density part, the rigid network of high melting point solid particles restricts the flow of the low melting point component and prevents their rearrangement. In a lower starting density part, the liquid is freer to flow and rearrange the solid particles to cause densification [26].

Although Cu particles are highly reflective, irregular or dendritic Cu powder is characterized by better absorption of the laser energy [11]. In addition, the particle size determines the minimum thickness of the laser sintered layer that controls the accuracy of the SLS parts. Therefore, a smaller particle size is preferred. However, fine powder particles are easily oxidized during the SLS process due to the presence of more surface area compared to larger particles.

Although the laser beam can produce sufficient energy to sinter high-melting point materials, the concentrated energy input causes a sharp temperature gradient in the powder layer. The residual stress induced causes distortion in the layer. To avoid this, the powder bed can be pre-heated to a particular temperature, depending on the materials [11].

5.3 Liquid phase sintering in SLS

5.3.1 Fundamentals of liquid-phase sintering

The powder binding mechanism is basically a liquid-phase sintering. In the two-phase powder system, when the laser scans the surface of the powder bed, the low melting point component melts and binds the high melting point powder together. Therefore liquid-phase sintering occurs, achieving high density and even nearly full density with short sintering time. The main advantage of liquid-phase sintering is the very fast initial binding. This binding is based on capillary forces, which can be very high. This transformation is faster than physical diffusion in solid state sintering.

The formation of liquid phase during SLS usually increases the sintering. The surface tension from the liquid phase is a significant factor in determining the sintering rate. For the SLS process, there are four main concerns:

(a) Amount of liquid formed during heating,
(b) Solid solubility in the liquid,
(c) Wettability of the liquid on the solid powder particles and
(d) Solid-phase diffusion in the liquid [24, 25].

Diffusion rates in liquids are relatively high, resulting in faster bonding and densification compared to equivalent sold-state sintering. For a normal liquid-phase sintering at the maximum temperature, full density can be achieved within duration of as short as 15 min.

When surface energies are dominant, liquid-phase sintering densification occurs. Initially, the mixed powder particles are heated to a temperature where liquid forms. Unlike normal sintering, there is solid state sintering during heating, which is also partly driven by the chemical concentration gradient in the microstructure. Since the laser processing is so fast, solid-state diffusion in general does not yield densification prior to the formation of the first liquid. Subsequent densification depends on the amount of liquid phase. Three stages of densification are encountered after the liquid is formed: rearrangement, solution precipitation, and the final-stage sintering.

If the liquid level is high, nearly full density can be achieved via rearrangement. On the other hand, at low liquid content, the solid skeleton inhibits densification, relying instead on solution-reprecipitation, where mass transport through the liquid controls densification. Residual final porosity can be eliminated by post solid-state sintering of the rigid solid skeleton.

As densification by rearrangement is slow, solubility and diffusivity effects dominate. This second stage is termed solution-precipitation. The solubility of a particle in its surrounding liquid varies inversely with the particle size. Small particles have higher energy and solubility than larger ones. The difference in solubility establishes a concentration gradient in the liquid. Material is transported from the small particles to the large particles by diffusion through the liquid. The process is termed Oswald ripening. The net result is a progressive growth of the larger grains at the expense of the smaller particles. Solution-precipitation not only contributes to grain coarsening but also to densification via grain shape accommodation, allowing a customized fitting of the growing grains to better fill the voids. The amount of liquid determines the diffusion distance and the necessary degree of green shape accommodation. The final stage of liquid-phase sintering is controlled by densification of the liquid structure. Densification is slow because of engagement of the rigid skeleton with the solid particles. Processes dominant in the final stage are also active earlier. However, because it is slow, they are not significant until late in the cycle. Microstructural coarsening continues and the residual pores are enlarged if they contain trapped gas, giving compact swelling. In general, properties of most liquid-phase sintered materials are degraded by prolonged final-stage sintering.

5.3.2 Binding mechanisms for liquid-phase sintering

When a liquid forms during laser heating, the microstructure consists of solid, liquid and vapor phases. The contact angle represents a metastable balance between the interfacial energy. For a liquid to wet a solid, the total free energy must decrease. The contrast between good wetting and poor wetting is illustrated in Fig. 5.7. Fig. 5.7 (a) shows a small contact angle between the liquid and solid, indicating good wetting, while in Fig. 5.7 (b) shows a large angle as the liquid cannot wet well with the solid.

The degree of wetting is characterized by the contact angle θ, which is the included angle in the liquid; its magnitude depends on the balance of three surface energies. For the point of solid-liquid-vapor contact, the three vectors balance in the horizontal direction,

$$\gamma_{SV} > \gamma_{SL} + \gamma_{LV} \cos\theta \tag{5.2}$$

where γ is the surface energy and the subscripts denotes the interfaces (SL = solid-liquid, LV = liquid-vapor, and SV = solid-vapor).

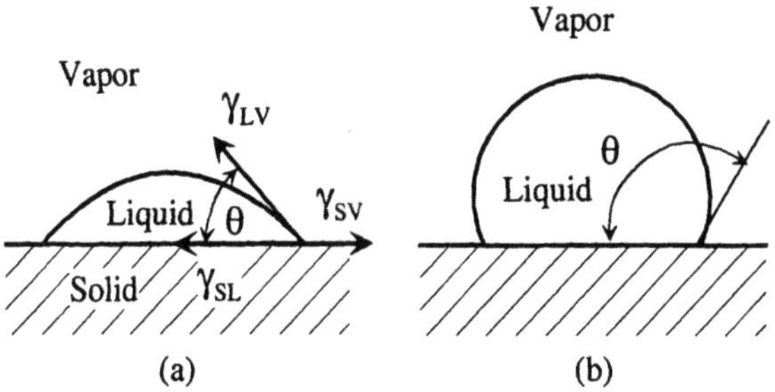

Fig. 5.7 Wetting between liquid and solid.

Note that the contact angle depends on the difference in interfacial energies. Surface impurities drastically alter the wetting behavior. That is why prevention of oxidation in SLS is extremely important. Similarly, higher temperatures reduce the contact angle because of increased solubility.

A wetting liquid occupies the lowest-energy position. Thus it preferentially flows to the smaller capillaries that have the highest energy per unit volume, leaving the larger pores empty. When there is insufficient liquid to fill the pores, the wetting liquid pulls the particles together to minimize the energy. This gives rise to rearrangement densification. In contrast, poor wetting means that the presence of the liquid on the solid surface is unfavorable. In liquid-phase sintering, poor wetting causes swelling of the compact and can possibly exclude liquid from the surface pores.

Like traditional liquid-phase sintering, the formation of liquid phase during laser sintering usually promotes sintering. The surface tension from the liquid phase is a significant factor in determining the sintering rate. Three main concerns are solid solubility in the liquid, wetting of the liquid onto the solid, and solid-phase diffusion in the liquid [24, 25]. As mentioned earlier, diffusion rates in liquids are relatively high, resulting in fasting bonding and densification compared to equivalent solid-state sintering. When surface energies are dominant, liquid-phase sintering densification occurs in stages, as can be seen in Fig. 5.8 that shows the amount of binding as a function of time. There are 3 different stages: rearrangement (melt and capillary penetration), solution reprecipitation (grain growth by migration of atoms) and a solid state sintering (final densification) [26]. If sufficient amount of liquid are formed with low viscosity to free flow, the final density at the end of the rearrangement stage is independent of the starting density. The amount of liquid is determined by laser power and scan speed. A very high sintering temperature can lead to a large amount of liquid phase with very low viscosity, which can result in loss of shape [2]. At low sintering temperatures, the amount of liquid phase formed is very small. The high viscosity also prevents any significant densification at the rearrangement stage [27-31]. The actual level of densification depends on the amount of liquid.

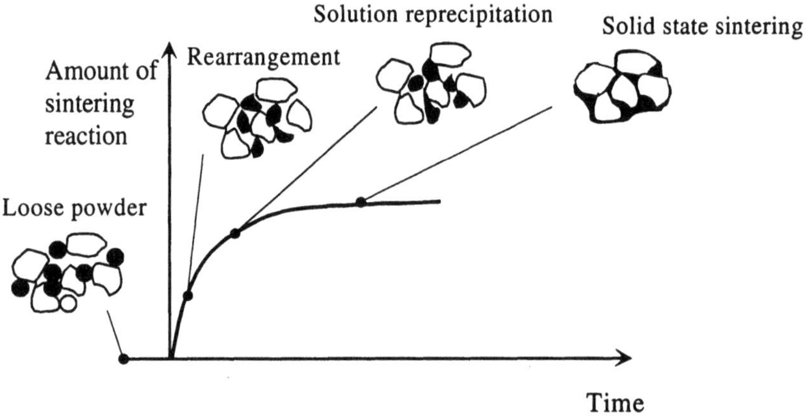

Fig. 5.8 Three stages during liquid-phase sintering process [26].

Since the duration of the laser heating is extremely short, typically between 0.5 and 25 ms [29], the thermally-induced binding reaction must be kinetically rapid. In the liquid-phase sintering, not like polymer binding mechanisms, melting is the key of binding. The flow of the liquid phase drives the formation of neck between two neighboring particles [30]. Several results have shown that the densification rate in the early stage of sintering often reveals an inverse viscosity dependence [32-34]. Sun *et al.* [35] has shown a sintering rate equation for SLS based on Frenkel's assumption and "shrinking cube" geometry (Fig. 5.9).

The sintering rate $\dot{x}$ is given as:

$$\dot{x} = -\frac{3(1-\rho)\pi\gamma a^2}{24\mu\rho^3 x^3}\left\{r-(1-\xi)x+\left[x-\left(\xi+\frac{1}{3}\right)r\right]\frac{9(x^2-r^2)}{18rx-12r^2}\right\} \quad (5.3)$$

where $2x$ is the cube size, ρ, the apparent density, γ, particle surface tension and ξ, a variable between 0 and 1. The amount of molten volume should be controlled in SLS. A viscosity μ of a solid-liquid mixture can be mathematically expressed by [36]:

$$\mu = \mu_0 [1 - (\Phi / \Phi_m)]^{-2} \tag{5.4}$$

where μ_0 is the viscosity of the pure liquid phase, Φ, the volume fraction of solid in liquid and Φ_m, the critical volume fraction above which the mixture has essentially infinite viscosity.

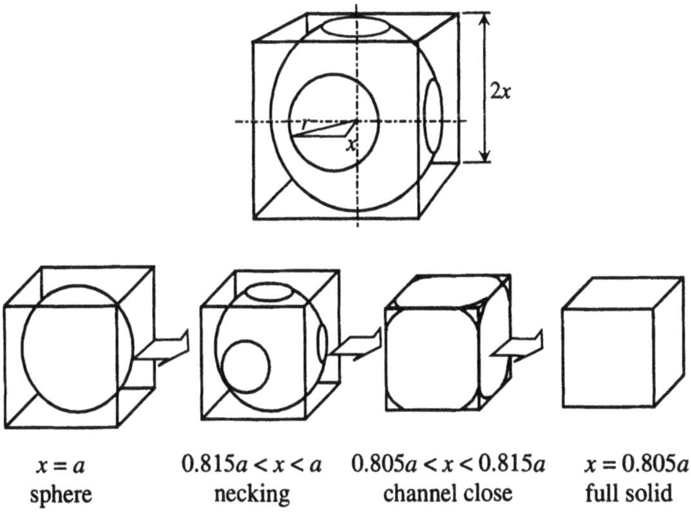

Fig. 5.9 Unit cell and process of densification [35].

Particle bonding is driven by the liquid viscosity μ_0, assuming the materials are suitably chosen for interparticle wetting to occur. The tendency for balling is controlled by μ, which is increased substantially by the retention of solid particles in the melt assemblage [29]. Agarwala [2] suggested that the viscosity of the molten liquid should be low enough such that it successfully surrounds the solid particles. However, if the viscosity is too low, there will not be enough liquid to cover the solid particles. In addition, if the liquid phase is too low, balling phenomenon may occur [29, 37-39]. Therefore the viscosity should also be high enough to prevent balling. The combination of low μ_0 and controlled μ can be achieved by the right solid-liquid ratio and by controlling the degree of superheating of the low melting powder [2].

Good wetting between the solid and the liquid phases can affect balling. It is well known that the wetting depends on a low solid-liquid surface energy in comparison to the sold-vapor and liquid-vapor surface energies. Wetting is aided by a chemical reaction at the solid liquid interface. The lack of chemical reaction is an indicator of poor wetting.

Good wetting may help to spread the liquid, which is very important and is a kinetic process associated with wetting. It is important to the early part of the liquid-phase sintering, since the liquid spreads and penetrates into pores between the powder particles, as fast as 2μm/s according to the traditional liquid-phase sintering. Solubility between the liquid and solid phase aids spreading, since spreading depends on a reduction in free energy, giving an increase in the liquid-vapor and solid-liquid surface areas and decrease in solid vapor surface area. This requires that

$$\gamma_{SV} > \gamma_{SL} + \gamma_{LV} \tag{5.5}$$

In other words, a liquid spreading over a solid is associated with a lower energy for the combined solid-liquid and liquid-vapor interfaces, as compared to the solid-vapor interface.

5.3.3 Influence of solubility

In liquid-phase sintering, there are the liquid solubility in solid and the solid solubility in liquid. A high liquid solubility in solid leads to a transient liquid phase, considerable processing sensitivity, and compact swelling during heating [40, 41]. On the other hand, solid solubility in liquid favors densification. Probably a solubility from 1 to 20 vol.% solid in liquid is typical.

Solid solubility in the liquid aids densification, as evident in the comparisons of sintered density versus time. A high solubility gives fuller densification (such as Fe-Cu). Note that solid dissociation in the liquid results in initial grain shrinkage, which helps rearrangement by separating the solid grains. A low solid solubility in the liquid, coupled to a high solubility of the liquid in the solid, results in swelling, since the volume of liquid decreases with time. Swelling can be reduced by making the initial structure more homogeneous through the use of small additive particle sizes or even coated powders. A homogeneous additive distribution before liquid formation gives more rapid densification during liquid-phase sintering.

Higher laser energy density is usually favorable for liquid-phase sintering, especially for low-solubility systems, because of increased solid solubility in the liquid and a greater quantity of liquid. For systems with a high solid-solubility in the solid additive, there may be some densification prior to formation of the first liquid [42, 43].

5.4 Commercial applications

5.4.1 Development of SLS

The first commercial metal-based rapid prototyping process was introduced in late 1995. Metal powders for the indirect process are typically coated or mixed with a small amount of polymer. In the SLS process, the metal particles are bound together by polymer-polymer bonds. This green part is processed in a high temperature furnace to remove the polymer and sinter the particles, creating metal-metal bonds. It is possible to produce full density parts using furnace sintering and other consolidation processes such as hot isostatic pressing (HIP). Typically, however, SLS green parts have a large amount of voids so that full density consolidation is accompanied by large shrinkage. To reduce total shrinkage, a partially SLS processed part can be infiltrated with a lower temperature material to produce a fully dense composite structure.

In the commercial SLS process, there are three distinct steps:

- A green part is made from carbon steel powder using the SLS process where only the polymer coating is melted.
- The polymer binder of the green part is removed by placing this green part in a reducing atmosphere furnace, and the metal powder particles are lightly sintered together.
- The lightly sintered part is infiltrated with copper.

In the commercial process, a furnace atmosphere of 30% hydrogen and 70% nitrogen is used. The hydrogen reduces oxides on the steel surface, while the nitrogen helps maintain carbon levels in the steel and prevents copper-induced swelling. The polymer is removed during two temperature holds: one at 350°C for five hours, and the second at 450°C for five hours. At these temperatures, the polymer decomposes and the degradated products are removed from the furnace by the gas mixture, which is circulated through the chamber at a fixed flow rate. Next, the sample is heated to a temperature of 1000°C and held for 8 hours. At this temperature, the steel powder begins to sinter at a significant rate. In the temperature region between polymer removal and sintering, the porous steel part is held together by friction between particles.

The University of Texas has developed a rapid tooling process for preparing mold inserts that are suitable for injection molding a limited quantity of polymeric materials [44-48]. In this process, SLS is used to form green mold cavity inserts from the metal powder that is coated with fusible thermoplastic binder. In subsequent steps, the binder is thermally removed and the metal powder is oxidized to form a porous metal/ceramic cavity that shows little shrinkage and generally excellent retention of geometry, relative to the green part. The cavity is then strengthened and sealed by infiltration and cured with an epoxy tooling resin. This mold-production process has been called rapid mold (RM). The advantages of the RM system are ease and speed of manufacture, low-cost post treatment in a simple air oven, and good hardness, thermal conductivity, and thermal expansivity. One interesting possibility is to use the RM process to prototype the plastic part, followed by using the RapidTool process to build the production mold using the same powder and model file of cavity part. The thermal conductivity of the RM material is lower than that of steel. However, it is still about seven times higher than the typical 0.2 W/m°C of unfilled polymeric materials [49]. Some estimation of the relative thermal performance of molds made of steel and of the RM material system has recently been made by solving the simultaneous partial differential equations that describe the transient heat transfer from polymer melt, through the mold material, to the cooling water [50].

5.4.2 DTM's RapidSteel and copper polyamide material

For tooling application, DTM's SLS system has equipped a roller type powder feeder and heat facility. The tooling materials available are the Copper Polyamide™ (a mixture of copper and nylon powder) and the RapidSteel™ (bronze-steel alloy). In the RapidSteel process, the laser beam scans over on the preheated powder layer so that only the polymer-binder is softened and melted. The metal powder is not melted by the laser during the sintering. After the SLS process, a green part is therefore formed. Post process of the green part should be carried out to increase its strength. To do this, the green part is heated in the furnace for several hours in order to remove the polymer, and is then infiltrated with bronze. Infiltration takes place by the capillary force. Furnace operation takes time and it is a tricky operation. The part is strengthened by the infiltration, though there may be unexpected amount of shrinkage during furnace operation. In the copper-polyamide process, only the shell with a certain thickness is generated, instead of sintering the whole solid block. No furnace operation is necessary and the inserts are machinable. Hence, the inserts have to be back-filled by aluminum epoxy resins. It can be fitted into a mould base for injection molding. The Copper Polyamide process allows short-run production (several hundred parts) in carbonate, polypropylene, and ABS for functional testing and concept evaluation. Specification of Sinterstation 2000 is given in Table 5.3.

Table 5.3 Specification of DTM's Sinterstation 2000 System

Model	Sinterstation 2000
Work volume	304.8 m diameter, 381 mm length
Maximum slice thickness	0.508 mm
Minimum slice thickness	0.0752 mm
Laser type	CO_2
Laser power, W	50
Unit size, XYZ (mm)	3022.6 X 1524.0 X 1930.4

5.4.3 EOS process

5.4.3.1 EOSINT M

EOSINT M is the first commercialized rapid tooling machine that utilizes the direct metal laser sintering (DMLS) technology. Electro Optics Systems

(EOS) of Munich in German has developed the machine since 1995 [21]. Fig. 5.10 shows the EOSINT M250 RP machine that uses 200 W infrared CO_2 laser to modulate the required laser energy. The laser spot size is about 300 μm and the laser cure width is about 600 μm. The laser scan speed is generally about 3 m/s and is reflected via rotary mirrors. The standard layer thickness is 50 μm, which defines the accuracy, built-time and stair-steps formation. Water is used for cooling the laser unit. Other specification is given in Table 5.4.

Fig. 5.10 EOSINT M 250 machine

There is a size limitation of the part to be built. Very large molds cannot be fabricated using the EOSINT M system. Increasing the built size requires the machine to be larger. The laser scan speed has to be regulated from the center to contour for a large slice area. The build time depends on the part size and design. No preheating is required in the EOSINT M system. This saves times to warm up and cool down the system.

Table 5.4 Specification of an EOSINT M 250 system

Building volume	250 x 250 x 150mm^3
Laser type	CO_2, 200 W
Laser scan speed	Up to 3 m/s
Building speed	2-15 mm^3/s (material dependent)
Layer thickness	0.05-0.1 mm
Electrical	400 V, 32 A
Dimensions (*B* x *D* x *H*)	1950 x 1100 x 1850 mm^3
Interface to CAD Standard	STL, CLI, VDA-FS, IGES, CATIA

In the EOSINT M system, parts are built on a metal base-plate, which is specially coated with bronze. This coated layer is needed to be evenly and uniformly distributed on the base-plate. Additionally this coated layer helps the first deposited powder layer to be strongly bonded to the base-plate under the laser sintering process. Hence, the quality of bronze coating may affect the bonding between the plate and the part. This is due to copper-rich material in the EOS material system. The thickness of the steel base-plate ranges from 10 to 15 mm.

The powder deposition system is an important design consideration. The quality of a deposited layer varies on the powder material and its size distribution because very fine powder is difficult to deposit. In EOSINT M, a blade is used. During the part building process, the blade brings the powder from the feed cartridge and spreads a new powder layer on the previous one. Then the blade returns back to the initial position, making the deposited layer flatter. To achieve a uniform quality part, an evenly flat layer is required in laser sintering. The blade speed ranges from 10 to 100 mm/s.

During the process, there is some gas emission that needs to be removed during sintering by an evacuation system with a charcoal filter at about $5m^3/h$. The gas evacuation also helps to reduce the formation of oxides in the sintered layers.

A SLICER software in a workstation performs the function of slicing of the part defined in the STL file by based on the user-defined orientation and slice thickness. The control computer of the machine gets the sliced files from the workstation, and arranges the orientations of the parts to be built. Several parts can be built in a single process. After the scanning parameters, such as scan speed, X and Y scales are set, The parts are built layer-by-layer. The machine can be interrupted when the feed-powder-cartridge needs to be refilled with powder.

5.4.3.2 EOS materials

The EOS system uses a pre-alloyed or multi phase powder instead of a single-phase pure metal powder. When the metal powder is exposed to the laser beam, its temperature is increased as laser energy is absorbed and sintering takes place at a certain temperature.

In solid state sintering, the sintering temperature $T_{sintering}$ should be at least one-half of the absolute melting temperature, i.e. ½ $T_{melting}$. The driving force for binding is the physical diffusion of metal atoms from one powder to another. By diffusion, the metal atoms move towards single powder particles (of the metal with higher melting point) forming 'sinter necks'. The slow phenomenon of this process is its major weakness for rapid tooling [26].

The EOS's sintering process adopts liquid-phase sintering. In laser melting, the metal particles with low melting point can be completely melted, forming a compact and nearly fully dense solid by adjusting the laser power. It is difficult to control the accuracy and surface finish of the laser-melted part.

Combined with a milling machine, laser melting can be applied for rapid tooling. No polymer-binder removal is required, and thus saving time.

5.4.3.3 EOSINT M 250 sintering process

In thc EOS process, copper-phosphide powder is used as a binder that melts at 660°C. When the laser beam exposes the powder, the particle melts, penetrates the surrounding cavities and wets the bronze/Ni particles [18]. Bronze may partially be melted at a higher temperature. Bronze and Ni homogenize at high temperatures, either by inter diffusion if the homogenization is between solid bronze and solid Ni, or by solution reprecipitation if the homogenization is between liquid bronze and solid Ni, to form a homogeneous solid solution of Cu in Ni [2]. Under a certain condition, the growth or swelling of the sintered part may occur [20]. By using the optimized parameters the typical shrinkage due to the liquid-phase sintering can be completely compensated by an expansion of the material caused by the diffusion of the component so that the material has no net volume change during the laser sintering process [21]. Due to the small laser focus of about 350μm and the high laser scan speed of 300 ~ 800 mm/s, the duration of heating is short enough to avoid oxidation even in a non-inert gas environment. In addition, the phosphor content inhibits oxidation of the powder mixture and also improves the wetting of the solid particles in the molten phase. The wetting and the diffusion are very important mechanisms to improve the mechanical properties of the laser sintered parts. It should also be noted that the most important variations are an increased content of liquid-phase, an enlarged surface area for the diffusion and an optimization of the laser and scan parameters.

The EOS process builds up parts on a piece of bronze-coated mild steel plate. To ensure good adhesion between the part and the plate, the plate is sometime exposed to the laser beam without powder for the first scan. The bronze-coating on the steel plate is melted completely and fused to the steel surface, ensuring a good adhesion.

The laser scanning of each layer includes contouring and hatching. There are two contouring stages. The first approximates the outlines with a certain offset. Then the interior area is hatched. The scan speed for hatching is slower than that for contouring. The second contouring stage exactly scans out the contour required of the sliced section. This procedure serves to generate a scanned section as accurately as possible.

5.4.3.4 Post process

After laser sintering, a stable but porous part with a density of about 70% of the theoretical density of the material is produced. The resulting part has a porosity of approximately 30% to 45%. Excess powders are removed with painting brushes and compressed air. The SLS processed parts can be sand blasted. At this stage, the parts can be machined by milling, grinding, drilling, etc. Finishing is necessary for the required surface quality.

To obtain a functional part, an infiltration process is necessary. High temperature epoxy, tin, silver, etc. may be used. However, the preferred infiltrate is epoxy resin because this material has minimum effect (shrinkage of about 0.1%) on the accuracy of the green part [21]. Although metal infiltration can improve the mechanical properties of the final part, the furnace process will also inevitably affect its accuracy. Note that Cu infiltration is not possible because its melting point (T_m=1083°C) is higher than the green part (T_m=900°C). Nickel plating can be applied to the EOS part in order to improve the surface finish, strength and wear resistance.

Because the EOSINT M process can build a complex geometry, manufacture of electrical discharge machine (EDM) electrodes using this process would be beneficial. In traditional, the complex shapes typically required by EDM electrodes are difficult and time consuming to produce. Since the EOSINT M material is a copper-rich material, thermal and electrical conductivities are very good. However, the porosity of the sintered parts may degrade the surface quality. Tool wear and shape deformation may be very high due to the presence of porosity in the parts. Ni-plating would improve the EOS electrodes for a better EDM process.

5.5 Metal powders for laser sintering

5.5.1 Development of laser sintering powder in general

The metallurgy of the SLS process is complex and therefore only limited kinds of metal combinations are suited. The material systems investigated to date include steels, Ni-base superalloys, Ti and its alloys, refractory metals, bronze-Ni and cermets, with bronze-Ni being the best combination so far.

Considerable effort has been devoted to the development of suitable powders. The first problem is the extremely short interaction time between the laser beam and powder particles being scanned. The loose powder has to be transformed into a rigid structure within a few milliseconds. Useful strength in an SLS-built part is accomplished by sintering with the use of a liquid phase. When the laser momentarily passes across the metal powder, the scanned powder is partially melted and it is this melted liquid metal that generates the inter-particle bonding. The second problem is to achieve dimensional stability during sintering. Because parts are built by sintering of successive layers generated from successive lines scanned by the laser, shrinkage would result in poor bonding between the lines and the layers.

Early attempts to SLS-process single-phase metals such as Cu, Pb, Zn or Sn were unsuccessful because of balling [29, 37-39]. When melting is induced by the laser beam, the molten powder quickly consolidates into spheres of diameter approximately equal to the laser beam diameter, rather than consolidating into the previous layer. This phenomenon is called "balling". At the lowest energy density, the powder does not sinter and remains loose. Once melting occurs, the resulting spheroidization is independent of the scanning rate and powder density. The spherical diameter increases with decreasing scan speed and increasing power. For this reason, a unique melting/solidification approach has been developed [51].

In an attempt to mollify the tendency to form spheres (spheroidization), a two-phase powder approach has been developed. Similar to liquid-phase sintering [25], the powder bed consists of at least two different powders with significantly disparate melting points. When the laser beam heats the powder bed locally, it only induces melting of the low melting point solid. The liquid formed wets and binds the high melting point solid. From the supersolidus liquid-phase sintering, the viscosity of a solid-liquid mixture is related to the viscosity of the liquid phase in the absence of any solid and the relative solid content. In this two-phase powder approach, particle bonding is driven by the liquid viscosity, assuming the materials are suitably chosen for interparticle wetting to occur. However, the tendency for balling is controlled by the viscosity of the solid-liquid mixture, which is increased substantially by the retention of solid particles in the melt assemblage. Therefore, by adjustment of the volume fraction of the high-melting-point phase, it becomes possible to induce binding without balling.

This two-phase powder approach has been used successfully in SLS processing of metals and ceramics. Examples of metallic and ceramic systems include Cu-Sn, Cu-(70Pb-30Sn) solder, bronze-Ni, Ni-Sn, alumina-ammonium dihydrogen phosphate, SiC-Si, WC-Co and alumina-boron oxide. A variation of this approach has been demonstrated in prealloyed 89Cu-11Sn bronze [52] by SLS processing in the supersolidus regime. The metal is only partially molten in a two-phase zone between the solidus and liquidus temperatures. The solid and liquid phases of the metal are in thermodynamic equilibrium, and this serves the same function as the binary powder approach.

A critical aspect of the two-phase solid-liquid approach to binding in SLS is the wetting of the solid phase by the liquid, which is time dependent, often to the order of minutes and hours. Transient wetting consideration is necessary in SLS due to the short time duration of the thermal exposure. A wetting balance may be used to assess the transient wetting behavior of superheated liquids on a solid substrate [53].

Binary metal systems for optimal density SLS processing should have the following characteristics:

(a) For the solid phase:

- fine particle size, high surface energy, good laser coupling;

(b) For the liquid phase:

- high surface energy, low viscosity, low solubility of the solid phase, no volatile components.

The melting points of the solid and liquid phase materials should be high so that substantial superheat can be achieved in the molten metal. The volume fraction of liquid must be greater than 30% for complete densification by infiltration wetting of interparticle regions of the solid phase. Liquid volume fraction exceeding 60% to 70% precludes either balling up due to the inadequacy of the solid framework to hold the liquid in the plane, or cracking due to the extensive shrinkage.

Binary metal powders investigated to date have involve the low-melting-point systems Cu-(70Pb-30Sn) solder and Cu-Sn. Starting powders included 99.0% Cu(-100mesh), 99.5%Sn(-100mesh) and a 70Pb-30Sn solder mixture (-100 mesh), with maximum particle size of 150um. Atmospheric control or flux additions are necessary to prevent surface oxidation and facilitate wetting. When a flux is applied, crystals of zinc chloride ($ZnCl_2$) are dissolved in alcohol and metallic powder is added [11]. The solution is evaporated under vacuum at temperatures between 60°C and 90°C, and the remaining powder is sieved to produce a flux-coated powder blend. The powder is SLS-processed in air or using a reducing atmosphere of $96N_2$-$4H_2$.

Density is an important factor in SLS. The density of an SLS-processed part is improved by decreasing the viscosity of the molten mass heated by the laser beam. For most powder systems, the relative solid content is limited to between approximately 0.25 and 0.6. With low solid content of below 0.25, the "balling" phenomenon may occur. With large solid content, insufficient bonding takes place due to the inadequate supply of low-melting-point phases, and the part does not hold well together.

5.5.2 Bronze-Ni powder

Prealloyed 90Cu-10Sn bronze mixed with Ni powder has been used successfully in SLS [3, 7, 54]. DTM (USA), Mareco (Netherlands) and EOS (German) have investigated this powder for a considerable period of time and successfully used it in injection molding. This System consists of bronze, nickel and copper phosphate and is a match to the Electrolux material [55] that is sometimes used in SLS systems [56]. Table 5.5 shows the properties of elemental Ni and Cu [57].

Fig. 5.11 shows the principle of micro soldering of the bronze-Ni powder by liquid-phase sintering. Ni has a high melting temperature of 1455°C while the melting temperature of the 90Cu-10Sn bronze is only about 940°C. During SLS, the copper phosphate serves as a flux. The laser beam only melts the bronze that fuses the Ni powder particles together. It is possible to further densify and homogenize the material by a post conventional sintering, since Cu, Sn and Ni form a solid solution. Some residual bronze is visible, but based on microstructural, X-ray and electron diffractions, the majority phase is the ternary solid solution. Supersolidus liquid-phase SLS of prealloyed, -150 mesh, 89Cu-11Sn bronze has been performed. This material begins to melt at approximately 850°C, but melting is not complete until a temperature in excess of 1000°C is reached. Within this temperature range, the material coexists as solid and tin-rich liquid.

Table 5.5 Properties of pure Ni and Cu

Physical property	Temperature	Copper	Nickel
Density (g/cm^3)	23°C	8.78	8.902
Thermal property	-	-	-
Melting point °C	-	1084.88	1453
Boiling point °C	-	2595	2730
Thermal conductivity (W/m·K)	350K	394	82.9 (373K)
	500K	388	-
Specific heat (J/kg·K)	-	386 (293K)	471 (373K)
Coefficient thermal expansion (μm/m·K)	0-100°C	17.4	13.3
Mechanical property			
Tensile strength (MPa)	-	209 (annealed)	317
Yield strength (0.2%) (MPa)	-	33.3 (annealed)	59
Elongation (%)	-	60 (annealed)	30
Elastic modulus	-	-	-
Tension (GPa)	-	128	207
Shear (GPa)	-	46.8	76
Compression (GPa)	-	140	207
Hardness	-	37 HRB	64 HV

5.5.3 Cu-Sn powder

The principle of design of Cu-Sn system is low melting temperature of Sn of 505.08K. During SLS, Sn is melted. Due to fast heat and cooling rates, molten Sn may only partially react with Cu to form solid solution and intermetallics that increase the strength of the Cu-Sn system. Mechanical properties vary with the amount of Sn. In general all the properties significantly increase with increasing Sn content. Fig. 5.12 shows the variation of properties of selectively sintered Cu-Sn system as a function of Sn [58]. The density of the SLS-built part increases because more melt forms during melting, resulting in a decrease in porosity. Several intermetallic compounds, such as $Cu_{31}Sn_8$, Cu_6Sn_5 and Cu_3Sn, can be formed due to inhomogeneous diffusion of Sn into Cu. Both increase in density and formation of intermetallics raise the impact toughness and hardness of the built part.

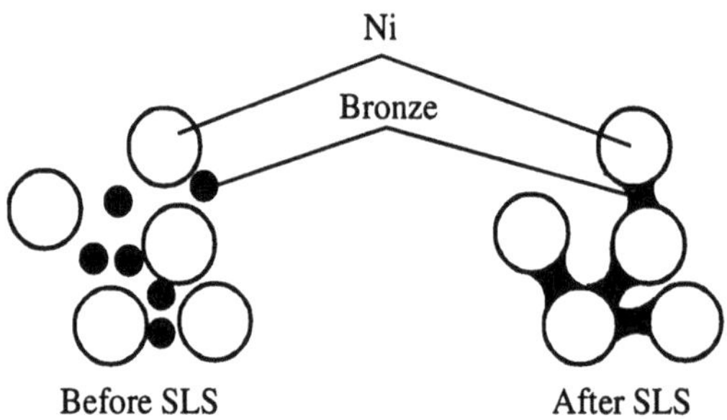

Fig. 5.11 Principle of micro soldering by liquid-phase sintering.

Although Sn can be fully melted, homogeneous solid solution may not be achieved due to the short period of heating. A number of intermediate compounds can be formed during the SLS process. Post processing should be carried out. The properties of the Cu and Sn powder are given in Table 5.6 [59].

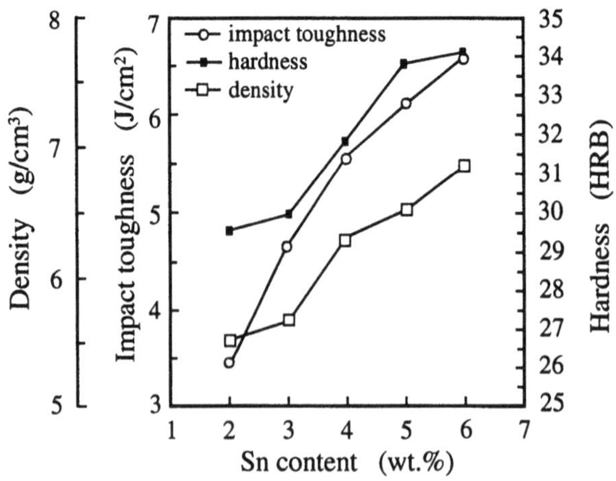

Fig. 5.12 Variation of properties with Sn (6wt.%C) [58].

Table 5.6 Composition and mechanical properties of Cu-Sn material

MPIF* density suffix	Composition %		Tensile strength (MPa)	Elongation in 25mm %	0.1% Compressive yield strength (MPa)	Density g/cm^3
	Sn	Cu				
N (I)	10	rem	55	1	48	5.6-6.0
R (III)	10	rem	96	1	76	6.4-6.8
S (IV)	10	rem	124	2.5	121	6.8-7.2

(* *Metal Powder Industries Federation*)

5.5.4 EOS powder

5.5.4.1 Powder

The EOS metal powder is a mixture of Ni, Cu, Cu-P and bronze developed and patented in the late 1980s by Electrolux Rapid Development, Finland for tooling production by pressureless furnace sintering [21]. This powder can be processed in a normal atmosphere without any technically complicated features like powder bed heating or additional powder bed compression devices [18]. Cu_3P acts as the low temperature binding agent. The particle size is approximately 30 ~ 45μm. The composition of the EOS powder system is given in Table 5.7. The morphology of the powder mixture is shown in Fig. 5.13. The particles with an irregular shape are Ni, and those with spherical shapes are bronze and copper-phosphide.

Table 5.7 Composition of EOS powder [60].

Material	Concentration
Cu	22-90
P	0.3-6
Sn	2-10
Ni	10-70

The EOS powder mixture is a Cu-rich system. The mixture is to be placed at a specially controlled area to prevent faster degrading due to humidity and surrounding temperature.

5.5.4.2 Role of Cu_3P in DMLS of Cu-Ni materials [3]

During sintering, Cu_3P melts and fills the surrounding cavities and wets the bronze/nickel particles. By using the correct parameters, the typical shrinkage due to the liquid-phase sintering can be completely compensated by an expansion of the material and diffusion of the components so that this material has no net volume change during the laser sintering process [21]. The influence of Cu_3P varying from 0wt.% to 10wt.% on Cu-Ni and bronze-Ni

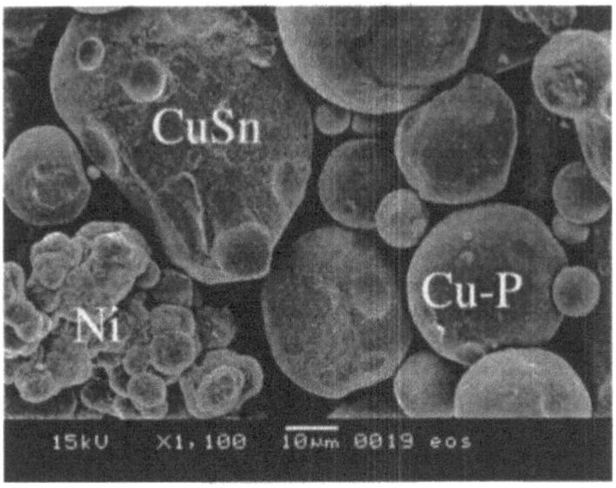

Fig. 5.13 Morphology of EOS powder particles.

powders has been investigated. It is shown that the present success of SLS of Cu-Ni powders is to a large extent, dependent upon the addition of Cu_3P. At the usual DMLS sintering temperatures (above the melting point of Cu, 1085°C), Cu_3P (with melting point at 1022°C) will melt and there is considerable solubility of P in the molten Cu. When Cu_3P melted, P is diluted by the Cu in bronze. Thus, after sintering, P is a solute in the Cu-P solution instead of forming Cu_3P. Smooth sintering layer (less balling) can be obtained when Cu_3P is added to the powder mixture. Therefore, the presence of P in the DMLS process may enhance sintering by improving the wettability of the Cu-Ni system by reducing surface tension of liquid Cu. Since the free energy of formation of P oxide is much smaller (more negative) than that of Cu and Ni, P may also act as localized oxygen "getter" and partially prevent the oxidations of Cu and Ni, and thus enhance sintering.

5.5.4.3 Hot isostatic pressing of DMLS bronze-Ni parts

Hot isostatic pressing (HIP) of the vacuum-sealed glass-encapsulated DMLS bronze-nickel samples can be done at high temperature from 750 to 900°C under pressure. The purpose of the HIP process is to close pores inside the SLS-processed parts.

5.5.5 DirectSteel 50V 1 (Steel-based powder)

EOS has recently introduced a new material for its EOSINT M process. It is a steel-based powder with some other metals but no organic components. The maximum particle size is 50 µm and can be used to build parts with very fine details. The metal powder exhibits negligible shrinkage during the sintering process and delivers very high accuracy. The new material does not require secondary infiltration and the processed part has a tensile strength of 500 N/mm^2 and Brinell hardness of 60-80. Seventy complex magnesium die-cast parts were produced from a DirectSteel die without any visible signs of wear in the tool. The building chamber has nitrogen-controlled atmosphere .

5.5.6 Tungsten carbide-cobalt powder

SLS of WC-Co-Ni system produced strong parts with rough surfaces. The wetting angle of WC by both liquid Co in hydrogen and liquid Ni in vacuum at 1500°C is 0°. However, the oxidation of Co and Ni causes poor wetting and thus balling. For a sintered layer, balling occurs at the top of the layer, while the bottom layer shows good wetting. At the bottom of the layers, WC wets well by metal matrix Co and Ni. This study implies that the oxidation of Co and Ni occurs only on the liquid surface during the SLS process.

Prealloyed tungsten carbide-cobalt powder has been SLS-processed with Co and Ni in the weight fraction 70 (88WC-12Co)-20Co-10Ni [62]. This was one of the first attempts to use RP techniques to directly generate a part from a conventional tool material. The part was fabricated in a high-temperature machine using 1.1KW CO_2 laser in nitrogen atmosphere. The surface texture of the SLS-processed part is in general rough, but extensive regions are almost fully dense. SLS of a nanocrystalline 50 alumina-50 cobalt systems has also been studied [63, 64]. The powder mixture is synthesized using sol-gel processing to produce aluminum and cobalt oxides 50 nm in diameter. The mixture is partially reduced by annealing in hydrogen at 1000°C to form an alumina/cobalt mixture. Following agglomeration in nitrogen at 1350°C, a promising, smooth single SLS-scanned layer of the powder was produced. Another nanocrystalline system studied was 20 nm to 50 nm tungsten carbide particulate SLS-processed after mixing with 15 percent cobalt [65]. The wear rate of the SLS processed material was approximately 5 times lower than that for traditional WC-15Co under identical test conditions.

5.5.7 Steel powder

In the selection of tool steels, the most important consideration is that the steel maintains integrity in shape and size over prolonged usage. The steel should therefore posses outstanding abrasion resistance, particularly when the molded part is to be made of either glass-filled thermoplastic or thermosets. They should be tough enough so that they will withstand extreme conditions of molding. For example, when a plastic piece is caught at the parting line, the steel will neither deform nor crack. There should be no noticeable degree of change in the properties of the steel at molding temperatures. The dimensional change after heat-treating should be negligible. It should have good corrosion resistance. After heat-treating to the desired hardness, it should be easy to polish the steel to a good mold finish. A good machinability index is an important factor especially when the configurations demand the use of small cutters.

5.5.7.1 Carbon steel

The optimized powder for the commercial process is 1080 carbon steel (RapidSteel metal) with a mean particle size of 55 μm [5]. Properties of three kinds powders are given in Table 5.8 [61]. The powder is uniformly coated with 0.8wt.% (approximately 5vol.%) polymer. The softening temperature of this polymer coating allows the steel powder to be processed in the SLS machine without heating the feed and part beds. The density of the powder in the part bed and the density of the green parts is 4.3g/cm^3, which is 55% of the theoretical density of the steel. There is negligible shrinkage during the SLS process. The strength of the green part is approximately 2.8MPa. This strength is sufficient for parts to be handled as well as for features as small as 0.5mm to be built and cleaned. Nearly all the unmelted coated powder is easily removed from the parts with a brush and can be completely recycled to make additional parts. The final part cleanup is accomplished outside the SLS machine with compressed air and small brushes.

Table 5.8. Composition of 1080, H13, and P20 steel (Balanced by Fe)

Powder	C	Mn	Si	S	Cr	Ni	Mo	Cu	V
1080	0.81	0.60-0.90	0.025	0.025	-	-	-	0.02	-
H13	0.32-0.45	0.20-0.50	0.80-1.20	<0.03	4.75-5.50	<0.30	1.10-1.75	-	0.80-1.20
P20	0.28-0.40	0.60-1.00	0.20-0.80	<0.03	1.40-2.00	-	0.30-0.55	-	-

In the portion of the furnace cycle after the coating softens, but before it is removed (between approximately 60°C and 400°C), gravity-induced creep of the lubricated particle structure can cause feature distortion as well as variable rates of setting in different regions of the part. This phenomenon can be eliminated by cross-linking the polymer binder. To accomplish this, the green parts are soaked in a dilute (3 wt.% ~ 10 wt.% solids) aqueous acrylic emulsion bath that contains a cross-liner. The saturated parts are then dried at approximately 50°C in an oven until all the water is nearly removed. It is also possible, in principle, to incorporate a cross-linker directly into the polymer coating and eliminate this step. The steel skeleton is infiltrated with copper at a temperature of 1120°C with a hold time of one hour. The infiltration is performed in a single cycle. To accomplish this, the green part is placed on a flat (less than 0.12mm distortion) alumina plate inside a graphite crucible.

The surface finish of the as-infiltrated parts is 10 μm R_a. To finish it as a mold part, it is necessary to correct critical dimension and polish display surfaces. Typically, metal-to-metal contact surfaces such as parting lines and shutoffs are machined. There is usually 0.5 μm of additional materials added to the parting line and shutoff surface in the CAD model of the mold to account for this correction. Mold polishing is performed with standard techniques. In some case, staircase effect on curved surfaces, which result from the layer-wise building method of the process, can be removed at the green part stage by light sanding. Additional mold finishing operations, such as chemical etching and plating, are also possible with the steel/copper composite material. The infiltrated part consists of 60% steel and 40% copper. The mechanical and thermal properties of this composite material are listed in Table 5.9 [66-72].

Table 5.9 Properties of the 1080 carbon steel infiltrated with copper [66-72]

Property	ASTM method	FeCu
0.2%Yield strength (MPa)	ASTM E8	255
Hardness (HRB)	-	75
Tensile strength (MPa)	ASTM E8	475
Young's modulus (MPa)	ASTM E8	210,000
Elongation (%)	ASTM E8	15
Thermal conductivity (W/m°C)	ASTM E457	91
Coefficient of thermal expansion (μm/m·°C)	ASTM E831 (-51°C to 232°C)	14.4

5.5.7.2 P20 and H13 steels

P20 is a chrome-molybdenum low alloy steel mostly in prehardened state to 300HB. It is suitable in this hardness for injection materials of the unfilled grades and where the tendency toward stringing and associated indentation is at a minimum. This tool steel can be carburized to a depth of about 1/16 in. with a resulting surface of 54-61 HRC [73]. Properties of P20 tool steel is shown in Table 5.10.

Table 5.10 Properties of the P20 tool steel [73]

Property	ASTM method	P20 steel
0.2%Yield strength (MPa)	ASTM E8	750
Hardness (HRB)	-	170
Tensile strength (MPa)	ASTM E8	950
Young's modulus (GPa)	ASTM E8	> 210
Elongation (%)	ASTM E8	20
Thermal conductivity (W/m°C)	ASTM E457	29
Coefficient of thermal expansion (μm/m/°C)	ASTM E831 (-51°C to 232°C)	1.7

H13 is a 5% chrome, high-alloy steel with all the properties required in plastic. It posses excellent abrasion resistance and toughness is not affected by temperatures. It can be heat-treated to a hardness of HRC 55-56. The change in dimensions and shape after heat-treating is insignificant. It is standard practice to plate thermoset molds with hard chrome, especially for the production of 10,000 or more per cavity [73].

5.5.7.3 Fe-based powder

Fe-based powder materials used for laser sintering are most often employed in prealloyed condition of Fe-Ni-Mo powder, which is used for the production of automobile parts. Prealloyed Fe-Ni-Mo powder with a 100 mesh is used for IVSI. WP200 elemental Fe powder with a parent density of 2.8-3.0 g/cm^3 and a particle size of -200 mesh is used for the WLMG of green compacts. FTD1 elemental Cu powder and C in the form of graphite (-300 mesh) are added to powder mixtures for fabricating two green compacts. The compositions of the two kinds of powder and their physical properties are given in Table 5.11 [22].

Table 5.11 Composition and properties of iron base powder [22]

Powder	Graphite	Fe-Ni-Mo	Cu	Fe	Hardness (HRB)	Density (g/cm^3)
IVSI	1.2	96.8	2	-	96-104	7.12-7.24
WLMG	1	-	1.5	Bal	91-95	7.10-7.19

5.5.8 Stainless steel

There are several kinds of stainless steel successfully used in SLS processing to date because of their perfect properties. The compositions of the common stainless steels used in SLS, including 304L, 316L and 17-4 PH, are listed in Table 5.12 [74]. Their mechanical and thermal properties under various conditions are listed in Table 5.13 [75].

5.5.8.1 316L Stainless steel powder

316L SS powder is used in indirect sintering applications [11, 76]. It is usually coated with a variety of polymer binders such as polystyrene and polyethyleneglycol.

5.5.8.2 17-4PH

All screening trials to date on SS 17-4PH have been conducted in a high vacuum environment with preheating. These trials are restricted to single layers. The initial trials indicate that highly dense, strong single layers can be made by direct SLS. The micrograph indicates that processing SS 17-4PH to high or full density is feasible. However, the bumpy structure indicates the individual scan and vectors are not completely bonded together, possibly due to inadequate laser energy input; it is believed that the material is nearly fully dense in the scanning direction [77, 78].

5.5.9 DTM powder

DTM's SLS process provides a very wide range of materials and applications due to the Sinterstation's ability to work with a variety of powdered materials. Current materials include DTM Polycarbonate, a family of DTM Nylons, DuraForm™ Polyamide, SOMOS 201 Elastomer, TrueForm™ Polymer investment casting material, SandFormä Zr & Si sand casting materials, RapidSteel™, and copper polyamide (PA) metal for tooling applications.

Table 5.12 Compositions of standard stainless steel

Powder	C	Mn	Si	Cr	Ni	P	S	Others
304L	0.03	2.00	1.00	16.0-18.0	10.0-14.0	0.045	0.03	2.0-3.0 Mo
306L	0.03	2.00	1.00	18.0-20.0	8.0-12.0	0.045	0.03	-
17-4 PH	0.07	1.00	1.00	15.5-17.5	3.0-5.0	0.04	0.03	3.0-5.0 Cu, 0.15-0.45Nb+Ta

Table5.13 Properties of stainless steel

Physical Property	Temperature °C	304L*	316L	17-4 PH***
Density (g/cm³)	-	7.8	6.6**	7.6
Tensile strength (MPa)	-	480	372**	1310
0.2% Yield strength (MPa)	-	170	275**	1170
Elongation (%)	-	40	4**	10
Reduction in area (%)	-	-	-	35
Hardness (HRB)	-	88	95*	40-48 HRC
Elastic modulus (GPa)	-	193	193*	196
Thermal property	-		(316L*)	
Melting range (°C)	-	1400-1450	1375-1400	1400-1440
Thermal conductivity (W/m·K)	100	16.2	16.2	18.3
	500	21.5	21.5	23.0
Specific heat (J/kg·K)	500	500	500	400
Coefficient of thermal expansion (μm/m·°C)	0 - 100	17.2	15.9	10.8
	0 - 315	17.8	16.2	11.6
	0 - 538	18.4	17.5	-

(** Annealed; ** P/M; *** H 900 treated*)

5.5.9.1 Copper polyamide

DTM's Cu PA material is a heat-resistant, thermally conductive composite of copper and plastic introduced specifically for use with DTM's Sinterstation™ system. The Cu polyamide material is processed in the Sinterstation at conditions similar to those of DTM's DuraForm polyamide material. The advantages of using Cu polyamide are both numerous and significant. Cu polyamide parts are machinable and finished easily with wet sanding. In addition, the heat resistance and thermal conductivity of the material are superior to most plastic tooling materials. The specification of Cu polyamide is given in Table 5.14.

Table 5.14 Specifications of copper polyamide.

Physical Property	Temperature °C	ASTM method	Built in Sinter-station 15 angle	Built in Sinter-station flat
Density (g/cm^3)	23	ASTM D792	3.45	3.45
Thermal property				
Thermal conductivity (W/m°C)	40	ASTM E1530	1.28	1.28
	150	ASTM E1530	0.92	0.92
Specific heat (J/g°C)	40	ASTM DSC	0.66	0.66
Specific heat (J/g°C)	150	ASTM DSC	0.87	0.87
Coefficient of thermal expansion (1/°C)	30-150	ASTM E831	92.6×10^{-6}	92.6×10^{-6}
Mechanical property				
Ultimate tensile strength (MPa)		ASTM D638	33.6	35.9
Flexural strength at 5% strain (MPa)		ASTM D790	53.5	58.3
Flexural modulus (MPa)		ASTM D790	3,068	3,223
Compressive strength at 0.2% offset (MPa)		ASTM D695	88	99
Hardness-Rockwell "D"		ASTM D2240	75	75
Surface finish, Ra (μm)				
As SLS processed			27.7	12.5
After finishing			5.2	2.5

5.5.9.2 RapidSteel 1.0 metal

RapidSteel 1.0 [79] is made up of low carbon steel powder with a mean particle size of 55 μm. The powder particles are coated with a polymer binder of low melting point, which allows the material to be processed using low power laser energy.

5.5.9.3 RapidSteel 2.0 metal

RapidSteel 2.0 is 316 stainless steel based powder system. The average particle size is about 34 μm [80]. The smaller size of the powder enables reduced layer thickness from 100 μm to 75 μm to be fabricated. RapidSteel 2.0 powder can be used to produce durable, full-density metal mold inserts using the SLS process. In the first step, the SLS process is used to produce "green parts" in which the steel particles are bound together by polymer. The green parts are then placed in a furnace where the polymer binder is removed and the steel particles are lightly sintered. The resulting "brown" part is approximately 60% dense. The brown part is returned to the furnace and infiltrated with bronze to produce a fully dense, steel/bronze composite tool. The resulting composite metal mold insert must then be finished and mounted in a mold base for molding production. The time required to prepare mold-ready inserts ranges from 5 to 15 working days. The specification of RapidSteel® 2.0 material is summarized in Table 5.15.

Table 5.15 Specification of RapidSteel® 2.0

Physical Property	Temperature °C	ASTM method	RapidSteel 2.0
Density (g/cm^3)	23	ASTM D792	7.5
Thermal properties			
Thermal conductivity (W/m°C)	100	ASTM E457	23
	200	ASTM E457	28
Specific heat (J/g°C)	100 - 150	ASTM DSC	339-418
Coefficient thermal expansion (1/°C) x 10^{-6}	-51 - 232	ASTM E831	14.6
Mechanical property			
Yield strength (0.2%) (MPa)		ASTM E8	413
Tensile strength (MPa)		ASTM E8	580
Elongation (%)		ASTM E8	0.9
Young modulus (GPa)		ASTM E8	263
Hardness-Rockwell "C"		ASTM E18	27

The resulting molds are 55% steel and 45% bronze, and have mechanical properties similar to P-20 steel and thermal properties superior to aluminum.

Depending on the type of plastic, force and temperature during injection molding, core and cavity sets created with RapidSteel 2.0 can produce in excess of 100,000 parts. In die casting applications, tools made with this process can produce several hundred die cast parts. RapidSteel 2.0 requires less finishing time due to the material's smaller particle size. This produces a smoother surface and allows parts to be built with a layer thickness as small as 75μm, which reduces the staircase effect. The new material system also provides improved consistency and accuracy, as a result of reduced process shrinkage. With earlier versions of RapidSteel 1.0, the total shrinkage is approximately 4% while the current RapidSteel 2.0 achieves a net dimensional change of only 0.2%. Applications can include mold inserts for injection molding prototype plastic parts, tooling for bridge and pre-production plastic injection molding, and mold inserts for die casting prototype non-ferrous metal parts. Key applications include consumer electronics, telecommunications, appliance manufacturing, healthcare, automotive, aerospace, and any other industry that must create prototype parts by plastic injection molding or metal die casting.

5.5.9.4 Process

Processes using DTM powder are illustrated in Fig. 5.14 [80]. For copper polyamide powder, no post processing is required while for RapidSteel powder after SLS, sintering and infiltration post processes are required to obtain full density. For the RapidSteel 1.0, there is an intermediate step of polymer infiltration followed by sintering and Cu infiltration.

Compared to RapidSteel 1.0, Cu infiltration is replaced by bronze infiltration for RapidSteel 2.0 powder. There is no polymer infiltration required. Sintering of green part is carried out at 1120°C for 3 hours. Infiltration for RapidSteel 2.0 green part is raised to 1050°C.

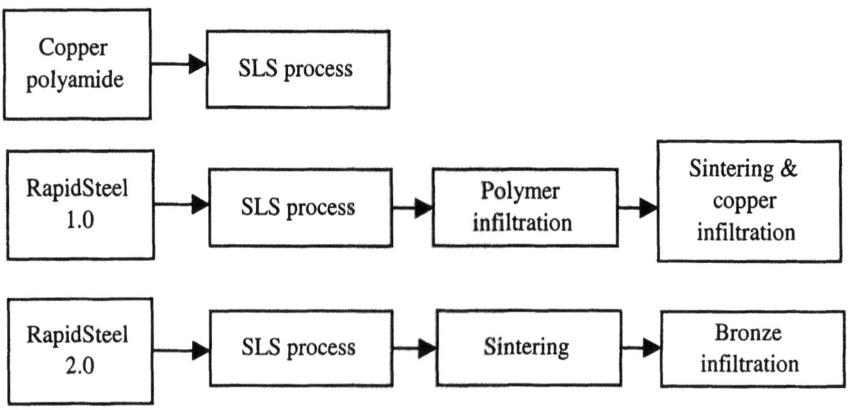

Fig. 5.14 Illustration of rapid prototyping process using DTM powders [80].

5.5.10 Mareco's R&D powder

Mareco has researched accuracy on laser sintered mold parts of approximately 50 μm. Because of the near zero shrinkage during the sintering, this accuracy is in most cases reproducible. In addition, the very fine powder allows for building in layers of 50 μm, resulting in high resolution with very fine details and hardly any disturbing staircase effect at angled sides. Mareco uses several master molds in which the sintered parts can be inserted. Therefore only limited machining is required for the sintered molds.

Quit often a rapid tool will be made out of a soft material. For this application Mareco uses several materials from which aluminum is in most cases a good choice. Special kinds of aluminium have been developed for use as mold plates. Aluminium has the advantage of easy machining and milling, resulting in much shorter lead times compared to hardened molds. In many cases, aluminium is a perfect soft tool material and very often machining of aluminium is the fastest way to produce a mold insert, especially when high-speed milling is used.

5.5.11 Nickel alloy

5.5.11.1 Inconel 625 superalloy

SLS processing of Inconel 625 has conducted in high vacuum with and without in-situ bake-out pre-processing [81]. Bake-out pre-processing helps to eliminate adsorbed contaminants from the powder particle surfaces, reduce oxide formation, and improve wetting and flow characteristics of the molten material during SLS. Surface oxidation of the single-layer coupon from this process can be much reduced when compared with the single-layer coupon produced without bake-out. This material exhibits overall reduced porosity in size and number than those processed using a higher temperature process.

The Argon-atomized Alloy 625 (equivalent to INCONEL 625) powder size is 16~44μm. The composition of Alloy 625 is shown in Table 5.16.

Table 5.16 Manufacturer's composition of Alloy 625 [81]

C	Si	Mn	S	Cr	Ni	Mo	Ti	Nb	Al	Cu	Fe
0.028	0.16	0.04	0.008	21.3	62.2	8.22	0.27	3.36	0.21	0.04	4.00

Integrally canned cylinders of Alloy 625 conforming to the geometry have been produced by SLS and post-processed to full density by HIP. Density before and after HIP has been evaluated by analysis of voids in a metallographic montage as well as by gas pycnomtry, showing 98.5% density

In this case, the entire diameter of the cylinder is consolidated to high density instead of producing an intermediate density core with a high-density skin. This specimen is post-processed by a HIP cycle of 3 hours at 1240°C and 155 MPa. This resulted in nearly full densification to 99.5%. Hardness value of SLS/HIP processed Alloy 625 is HRC52. There is no significant difference between hardness of 98.5% dense SLS processed material and 99.5% dense HIP post-processed material [77 - 82]. The properties of Alloy 625 are listed in Table 5.17.

Table 5.17 Mechanical properties of Alloy 625

Property		Temperature (°C)	Alloy 625
Density (g/cm^3)		21	8.4
Tensile strength (MPa)		21	855
		540	745
0.2%Yield strength (MPa)		21	490
		540	405
Elongation (%)		21	50
		540	50
Hardness (HRC)		-	52
Elastic modulus (GPa)	Tension	21	208
	Shear	21	81
Impact strength (J)		29	66
Melting point (°C)		-	1290

5.5.12 Titanium alloy

At University of Texas, Austin, Ti-6Al-4V parts have been fabricated using SLS on a blended elemental powder in high vacuum, 400°C preheat and bake-out pre-processing [78]. At lower laser energy density, single layers with relatively good surface texture and integrity can be formed. However, due to low laser power, these single layers exhibit a high degree of porosity. Increasing the laser energy density in order to create dense single layers results in severe balling or agglomeration of the molten material. Balling and agglomeration are expected to be significantly alleviated under cleaner atmosphere (high vacuum), higher preheat temperatures and optimization of laser processing parameters [78].

The powder size of Ti-6Al-4V produced by the PREP method is 37~74 μm. The composition and properties of Ti-6Al-4V are given in Tables 5.18 and 5.19.

Table 5.18 Manufacturer's composition of PREP Ti-6Al-4V [81]

Al	V	Fe	O	C	N	H	Y	Ti
6.35	4.19	0.19	0.19	0.02	0.01	<0.01	<0.001	Bal

Integrally canned cylinders of PREP Ti-6Al-4V conforming to the geometry have been produced by SLS and post-processed to full density by HIP. While the side-walls and end-caps of the Ti-6Al-4V powder cylinder are SLS-processed to full density, the interior of the part is left unprocessed. However, necking between powder particles in the core region occurred due to local heat transfer from the skin region, resulting in a core density of about 60% theoretical density. The HIP cycle for this specimen consists of 4 hours at 925°C and 93MPa. Fig. 5.15 shows the overall microstructure of HIPed SLS-processed Ti-6Al-4V. It shows clearly the difference between the skin and core. More detailed microstructure is given in Figs. 5.16 (a) and (b), which reveal the microstructures of fully dense skin and HIP-processed fully dense core [81]. The core region resulting from coalescence of prior particle grain boundaries during HIP differs from that of the fully melted and resolidified skin region. The core region displays a Widmanstatten or basket weave microstructure. Metallographic evaluation of the HIP post-processed specimen reveals a mean residual porosity of only 0.1% and homogenization between the core and skin regions. Hardness values of SLS/HIP-processed Ti-6Al-4V specimens indicate that the SLS/HIP-processed material compares well with the conventionally processed material [77-83].

Table 5.19 Mechanical properties of sintered Ti-6Al-4V

Property	Temperature (°C)	Ti-6Al-4V
Density (g/cm^3)	21	4.43
Tensile strength (MPa)	-	162
0.2%Yield strength (MPa)	-	884.6
Elongation (%)	21	23
Hardness (HRC)	-	36
Elastic modulus (GPa)	21	110
Impact strength (J)	90	29.8
Reduction in area (%)	-	23
Thermal property		
Melting point (°C)	-	1604
Boiling point (°C)	-	1660
Thermal conductivity (W/m·K)	20	6.6
	90	7.3
Specific heat (J/Kg·K)	20	580
	205	610
	425	670
Coefficient of thermal expansion (μm/m·K)	0 - 200	8.6 - 9.7
	200 - 400	9.5 - 10.3
	400 - 600	9.7 - 10.7

Fig. 5.17 shows a subscale version of the guidance section housing for the AIM-9 Sidewinder missile made by SLS/HIP process [83]. Fully dense

microstructure can be obtained and the microstructure of the HIPed one is consistent with the cast Ti-6Al-4V.

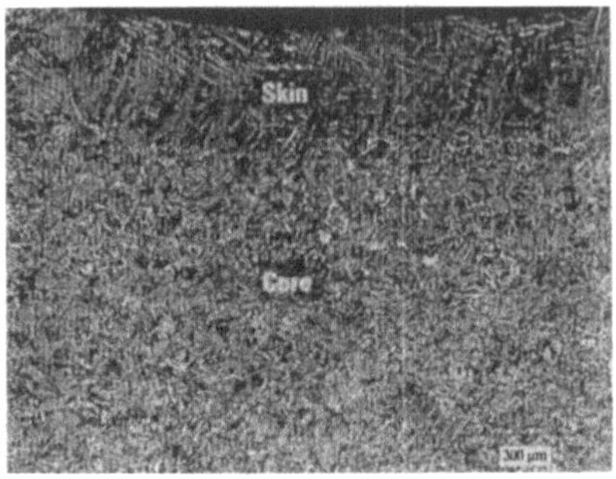

Fig. 5.15 Low magnification of Ti-6Al-4V microstructure of skin and core [81].

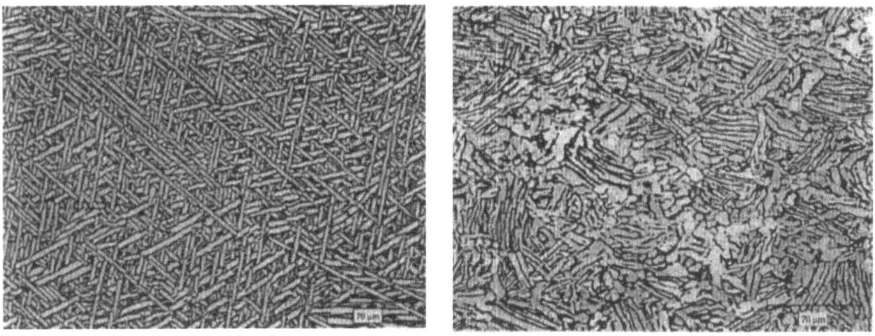

(a) (b)

Fig. 5.16 Microstructure of Ti-6Al-4V SLS processed structures: (a) fully densed skin and (b) fully densed core after HIP process [81].

Fig. 5.17 SLS processed Ti-6Al-4V [83].

5.6 Densification

A nearly full relative density is one of the most important properties required for tools, and molds in particular. For instance, porosity can lead to tool failure because of low mechanical resistance, elastic limit, Young's modulus and toughness. It also affects the surface quality. In the case of injection molding, porosity may also influence the dimensional accuracy of the mold, because the material may be further densified during the injection and hold pressure after a few hundred cycles.

SLS does not produce full density parts because in this case, the powder cannot be compacted before the treatment. During the SLS process, loose powder bed density and wetting properties influence the degree of porosity. After binder burnout, the brown body is porous and brittle, and needs further densification either by conventional sintering and/or infiltration. However, infiltration does not allow complete densification of the part. It appears to only penetrate a finite depth below the surface. With parts for injection molds, insufficiently infiltrated cooling channels may leak. Dissolution into the coolant of some powder components may also happen [16]. Sometimes the non-metallic infiltrant phase within the tool is likely to induce additional internal stresses because of inhomogeneity.

In SLS, the laser exposure strategies play a key role in the part quality. If a high relative density is required, the build time is very long because the successive beam paths have to be very close to each other and the scan velocity must be low. Hence, advanced exposure strategies, such as the skin and core strategy, have been developed by manufacturers to reduce the fabrication time while maintaining required strength.

Since density is a major parameter affecting tool strength, durability and surface quality, it is highly important that the methods used must be capable of producing almost fully dense parts for as a wide range of materials as possible.

5.6.1 Post sintering

Sintering is a post process of SLS. Post sintering aims to increase the density and homogenize the composition of SLS-processed parts. The parts fabricated by laser sintering generally suffer from two main types of defects: high porosity in the structure and inhomogeneous composition. The cause of porosity is due to partial melting of the powder during the SLS process while the imnhomogeneity of the composition and, hence structure, is due to the short sintering duration during the liquid-phase sintering in SLS. The inhomogeneity of composition can be seen in Fig. 5.18, which clearly reveals the change in microstructure across Cu-Sn interface after laser sintering. Therefore, further densification and homogenization are important.

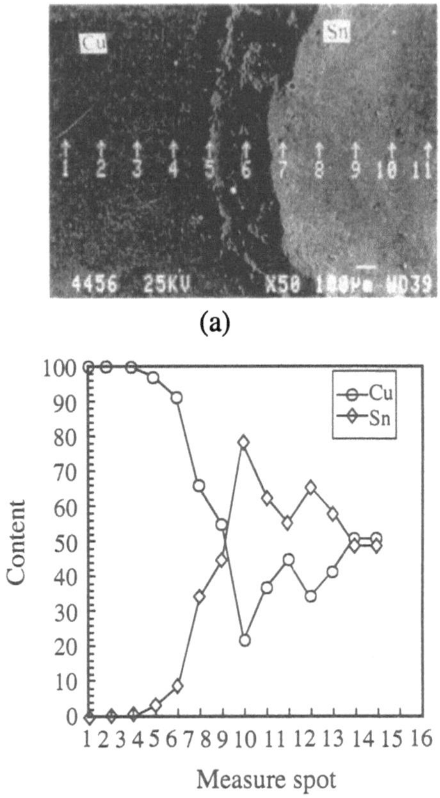

(b)

Fig. 5.18 (a) Microstructure of laser sintered Cu and Sn showing gradient of microstructure, (b) variation of Cu and Sn contents in interface [58].

The sintering temperature is determined by the types of material systems. In general the sintering temperature is above half of the absolute melting temperature of the material to be sintered. During high-temperature sintering, the amount of pores decreases with the sintering duration. At the same time, the high and low temperature components may react each other to form a homogeneous structure. Full densification using conventional sintering process is almost impossible, but the density of the sintered part can be dramatically increased. Fig. 5.19 shows schematically three parts sintered at different conditions. All parts show an increase in density with sintering duration due to the driving force of surface energy. The part sintered in the vacuum condition has a higher density compared to that sintered in argon condition. In vacuum, the gases in the pores are evacuated facilitating the closing of the pores. As the sintering temperature is raised, the driving force for the reduction in surface area rapidly increases, leading to fast densification, as compared to low-temperature sintering. However, it should be noted that high-temperature sintering may lead to increase in the grain size, which may degrade the properties of the parts.

5.6.2 Infiltration

Conventional sintering may give rise to large shrinkage due to reduction in pore size and volume. The change in the volume can be reduced by liquid infiltration process. The processes of liquid infiltration can be divided into three different categories: pressureless, vacuum, and pressure infiltration. An infiltration process is associated to capillary effects causing flow of a viscous liquid into the SLS parts [84].

The process of an infiltration is schematically shown in Fig. 5.20, where a SLS part is placed together with the infiltration materials (Fig. 5.20 (a)). The SLS part and the infiltration materials are heated to a temperature at which only the infiltration metal is melted. The molten metal gradually fills the pores by capillary forces (Fig. 5.20 (b)). If the infiltration material has low viscosity and good wetting with the part, transportation of the liquid material to the SLS part may take only a short time.

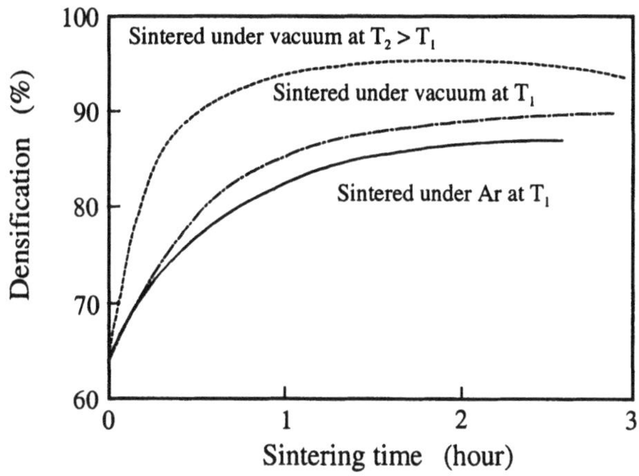

Fig. 5.19 Schematic illustration of change in density of sintered SLS parts in different sintering condition.

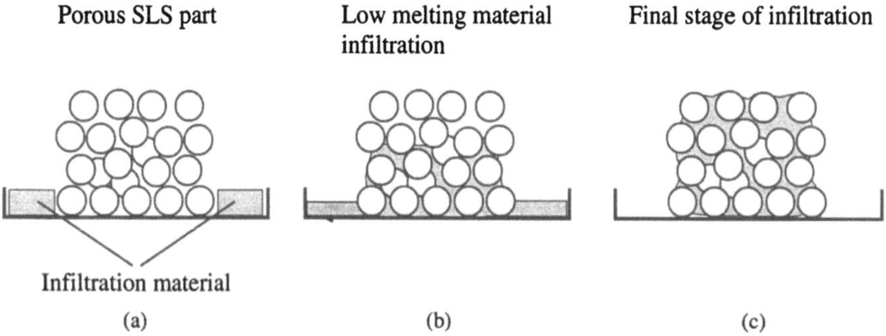

Fig. 5.20 Infiltration of part. (a) before infiltration, (b) during infiltration and (c) completion of infiltration.

Infiltration requires the pore to be open and interconnected. The part must have at least 10% porosity. Ideally, the liquid should not form intermediate compounds with the part since the formation of intermediate compounds may block the infiltration path. However, the liquid should have little solubility with the solid part [25].

The driving force of the infiltration process is surface energy, which can be determined by the measurement of the contact angle between the liquid and the solid. The contact angle in turn determines the infiltration behaviour (and liquid-phase sintering).

When a liquid droplet is in contact with a solid, three surface energies, solid-vapor energy γ_{SL}, solid-vapor energy γ_{SV}, and liquid-vapor energy γ_{LV}, should be in equilibrium condition, as shown in Fig. 5.7.

The mathematic expression of the three surface energies at equilibrium can be written as:

$$\gamma_{SV} = \gamma_{SL} + \gamma_{LV} \cos\theta \tag{5.6}$$

Infiltration can be modeled as a capillary pipe of diameter *d* that is equal to the pore size (Fig. 5.21).

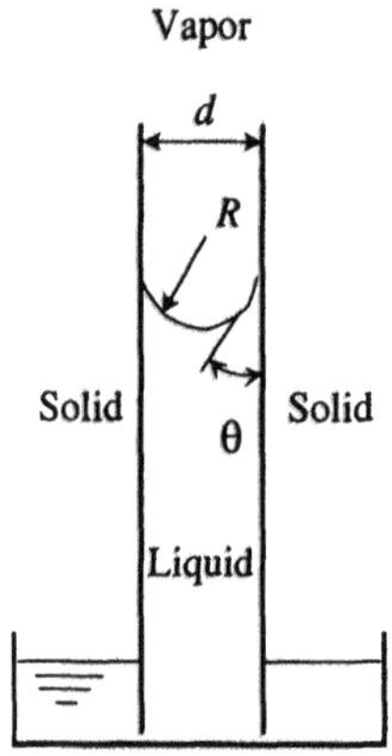

Fig. 5.21 Capillary force in an infiltration.

The depth of infiltration at the initial stage can be estimated by [87]:

$$h = (2/\pi)\left[\frac{d\,t\,\gamma_{LV}\cos\theta}{4\mu}\right]^{1/2} \tag{5.7}$$

where *h* is the depth, *t*, the time and μ, viscosity. Since the rate of infiltration is thermally activated, the depth increases with an exponential dependence on the temperature.

The pore size is also critical since the capillary attraction is increased by a smaller pore diameter. As mentioned earlier, the attractive force between two powder particles with a wetting liquid causes rearrangement, densification and contact flattening.

5.6.2.1 Cu and bronze infiltration

A number of metals for infiltration has been investigated. These include Ag, Sn, Pb, Cu and bronze among which Cu and bronze have demonstrated better infiltration and mechanical properties.

Fig. 5.22 shows the furnace cycle used to debind, sinter and Cu infiltrate green parts. Since oxide layer on the surface of powder particle can dramatically reduce wetting, a furnace atmosphere of 30% hydrogen and 70% nitrogen is used in commercial process. The hydrogen can reduce oxides on the surface of powder while the nitrogen helps maintain carbon levels in the steel and prevent Cu-induced swelling [79, 85]. For indirect SLS parts, the polymer in the part is first by holding it at 350 and 450°C for 5 hours. The powder starts to bond together at 1000°C. This sintering process will take about 8 hours. As long as the powder particles form a skeleton, infiltration of Cu can be carried out at 1120°C. Since liquid Cu has small contact angle with Fe (contact angle = 0°), infiltration takes place by capillary force without external pressure. The period of infiltration is short, about 1 hour. The Cu infiltration contains about 2wt.%Fe in order to prevent erosion of the iron part due to solution of Fe into liquid Cu [67]. If bronze is used, such as infiltration of RapidSteel 2.0, the infiltration temperature is set at 1050°C and held for 2 hours [80].

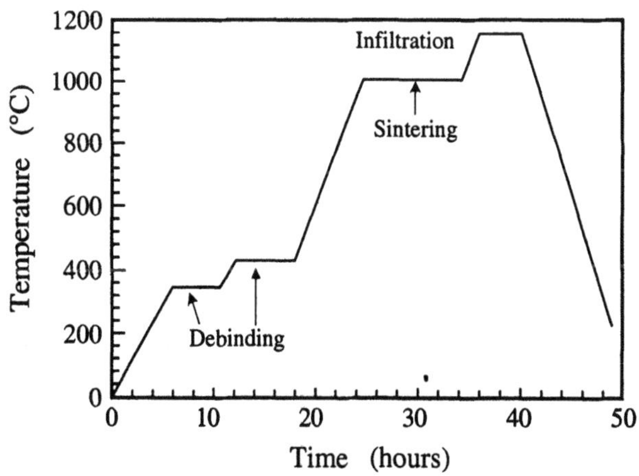

Fig. 5.22 Furnace cycle of copper infiltration [79].

5.6.2.2 Epoxy infiltration

Low melting point metal and epoxy resin can be used for infiltration. Epoxy infiltration is a better choice because it is a simple process and easy to control. Epoxy infiltration also makes part smoother and stronger. The process includes part preheating, infiltration, brushing, degassing and baking. It is estimated that about 30 % of the sintered part are cavities that can be filled with epoxy after infiltration.

A high-temperature epoxy resin for infiltration has been investigated. Before infiltration, the part has to be cleaned, without any dust or loose powder.

The surfaces of the parts are brushed with epoxy. Epoxy is just poured into the holes where brushing cannot be applied. By the capillary effect, the epoxy fills the pores of the part. The parts are then placed in a vacuum desiccator for degassing to evacuate the residual air present in the parts, thus allowing the epoxy resin to fill further. After removing from the vacuum desiccator, the parts are heated at 50°C for 10 min. Then, epoxy is applied again. The process of brushing, degassing and heating continue until the surfaces of the parts become wet even after drying in the oven. In the final stage, the parts are cured in the oven at 160°C for two hours. The epoxy can also be infiltrated from bottom to top by capillary forces; it can also be poured from the top of the parts; or the parts can be dipped into the resin bath.

The infiltration epoxy is a mixture of two compounds, of which compound A is a mixture of resin and leveling agent, and compound B is a hardener. Compounds A and B should be thoroughly mixed to a ratio of 1:0.2546 at about 50°C.

Table 5.20 shows the mechanical properties of infiltrated Electrolux material.

Table 5.20 Mechanical property of infiltrated Electrolux material [6]

Property	Electrolux material epoxy-infiltrated
Tensile strength (N/mm^2)	157
Bending strength (N/mm^2)	372
Brinell hardness	96 ~ 108

5.6.3 Hot isostatic pressing

HIP was initially used for diffusion bonding of nuclear fuel elements. It has been used in the consolidation of power metallurgy. This technique can also be used in the consolidation process of mechanically alloyed materials. HIP normally involves several components, such as pressure vessel, heating element, compressor and temperature and pressure controllers. In general, HIP may be divided into two different categories based on different pressing methods. The first category is called capsule method. In this method, the powder is enclosed in a container (capsule) using a suitable canning system. Sometimes it is called encapsulated HIP, which is the only processing route that allows pressure-added densification over the complete range from green compact density to theoretical density. The encapsulation containers are normally made of low carbon steel, stainless steel, Ti, Ni and Cu. It must be noted that thermodynamic reduction of sheet metal grain boundaries and weld metal oxides may occur in the high-pressure Ar that may subsequently lead to leakage [86]. Pyrex, Vycor and quartz glass containers are also very often employed [88, 89]. The container should be well vacuumed to at least $10^{-2} \sim 10^{-4}$ Torr. Sometimes the vacuum process must be carried out at an elevated temperature. Under such condition, the container should be hermetically sealed by welding to make the container gas tight. For glass containers which can be produced by glass blowing or casting a slurry of ground glass into the required form, they will become soft and continuously decrease in viscosity with increasing temperature. Although the glass containers will become soft, they may be used at as high a temperature as 1,500 to 2,000°C where sheet metal container cannot be employed [88]. Sheet metal containers may be removed by machining or chemical milling while glass containers tend to break from the surface after HIP. The common process for HIP is given in Fig. 5.23.

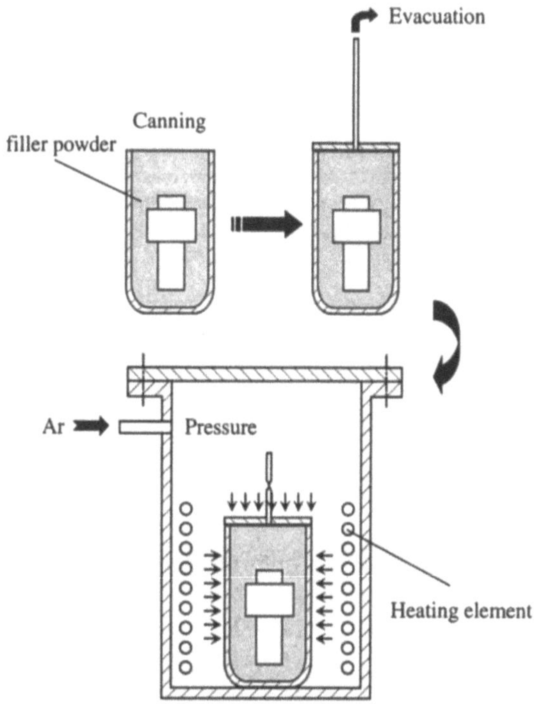

Fig. 5.23 Process route of HIP.

Fig. 5.24 shows an example of the change in density of bronze-nickel parts that have been HIPed at 124MPa at different temperature and duration [2]. The density of the HIPed part increases with both temperature and sintering duration. Compared to HIPing duration, HIPing temperature is a predominant factor affecting density of the parts. Table 5.21 lists some of HIPing parameters for a number of SLS-processed materials.

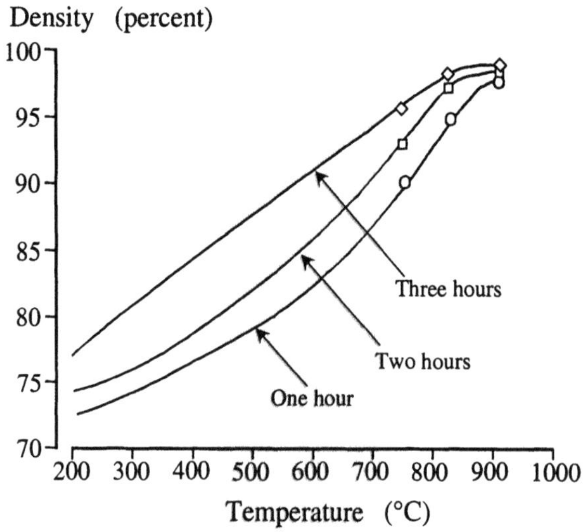

Fig. 5.24 Change in density of SLS-processed bronze-nickel parts as a function of HIP temperature and time at a constant pressure of 124MPa [2].

Table 5.21 List of containerless HIPped materials and operation parameters

Containerlss HIPped material	Pressure (MPa)	Temperature (°C)	HIP duration (h)
Ti-6Al-4V [81]	155	1240	3
Ti-6Al-4V [83]	103	925	4
Inconel 625[77]	155	1240	3
17-4PH SS [77]	-	-	-
Molybdenum [77]	-	-	-
Bronze-Ni [2]	124	900	1 ~ 3

HIP using cans sometime lead to contamination due to intereaction between the can and the SLS-processed parts or between filler powder and the SLS-processed parts. Therefore containerless HIP process has been developed [81, 90]. This process is also called in-situ encapsulation. In in-situ encapsulation, the interior portion of the part is scanned with a relatively low laser power in a vacuum environment. Only a porous interior structure with 65% or higher density can be formed. A high-power laser is used to scan most of the outer layer forming fully dense case as schematically illustrated in Fig. 5.25. Fully densified case should not have any leak. The in-situ SLS-processed parts can be directly post-processed by containless HIP.

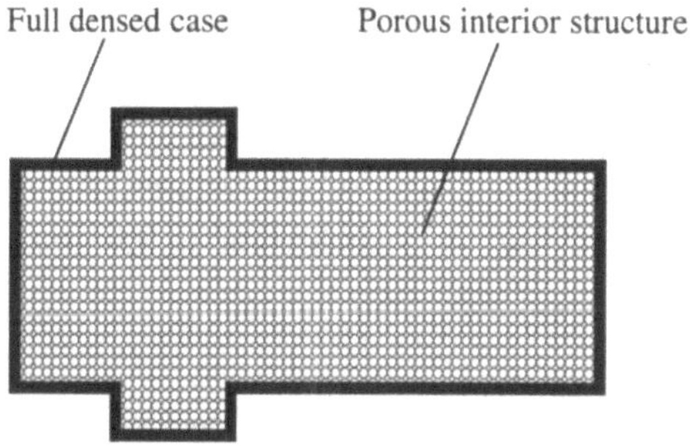

Fig. 5.25 In-situ encapsulated SLS part.

The comparison of microstructure of SLS/HIP-processed and conventionally processed Ti-6Al-4V is shown in Fig. 5.26. Although there is some difference in microstructure between the core and the skin of the processed Ti-6Al-4V due to the melting of the skin region in the SLS process, the bulk microstructure of the SLS/HIP-processed part is the same as that of the conventionally processed part that is encapsulated in glass and post-processed using the same HIP cycle.

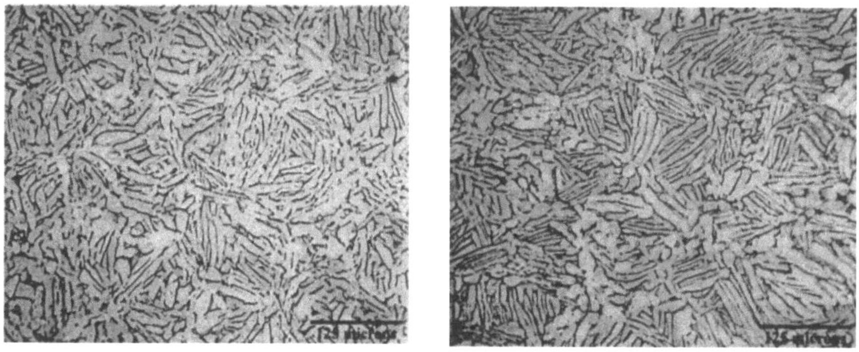

(a) (b)

Fig. 5.26 Comparison of microstructure of glass encapsulation HIP processed Ti-6Al-4V (a) and SLS/HIP processed Ti-6Al-4V (b) [83].

5.7 Mechanical property

It is important to design a process and material system in order to achieve maximum density of the sintered part. Since laser-processed parts in general consist of porosity, the strength of the part is proportional to its porosity [25, 91]:

$$\sigma_p = C\sigma_0 D^m \tag{5.8}$$

where σ_0 is the strength of fully dense material, D, relative density, C, empirical constant and m, constant varying between 3 and 6 for metal-ceramic materials.

Fig. 5.27 shows a very good agreement of the predicted strength using Eq. (5.8) with the measured data of the SLS-processed bronze-Ni part without post processing. The constant C and m are respectively equal to 590.8 MPa and 9.43. It can be seen that it is important to minimize the existence of porosity in the parts.

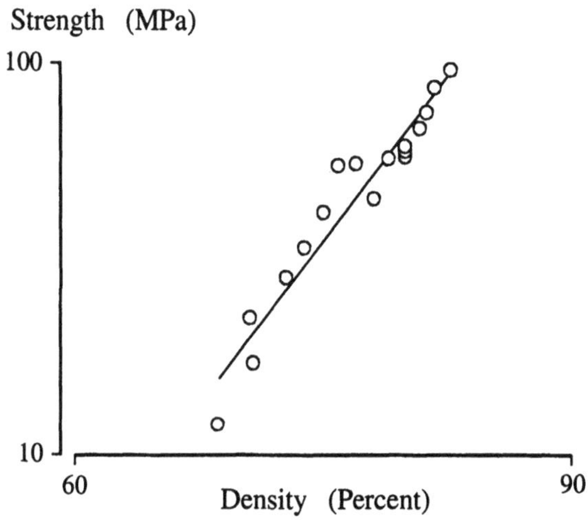

Fig. 5.27 Strength as a function of density of predicted using Eq.5.8 (the solid-line) and of SLS-processed bronze-nickel parts without post-processing (open circle) [2].

5.8 Summary

Indirect and direct selective laser sintering have been developed and commercialized since the discovery of laser-based rapid protyping by the University of Texas at Austin. The selective laser sintering process in general consists of two stages, namely laser sintering and densification processes. To meet specific application requirements, different material systems, such as RapidSteel, EOS powder, have been developed. By changing the chemical composition of the materials and processing parameters, shrinkage can be nearly fully compensated. Mechanical properties of some of the laser-sintered parts are comparable to those fabricated via conventional processes.

5.9 References

1. N.K. Tolochko, T. Laoui, Y.V. Khlopkov, S.E. Mozzharov, V.I. Titov and M.B. Ignatiev, *RP. J.*, Vol. 6 (2000), 155.
2. M. Agarwala, D. Bourell, J. Beaman, H. Marcus and J. Barlow, *RP. J.*, Vol. 1, (2) (1995), 36.
3. G. Zong, Y. Wu, N. Tran, I. Lee, D.L. Bourell and J.J. Beaman, *Proc. Solid Freeform Fabr. Symp.*, Ed. H.L. Marcus, J.J. Beaman, J.W. Barlow, D.L. Bourell and R.H. Crawford ,The University of Texas, Austin, Texas, 3-5 Aug. 1992, 72.
4. C.R. Deckard, *Method and Apparatus for Producing Parts by Selective Sintering*, U.S. Patent 4,863,538, Oct. 17, 1986).
5. H.L. Marcus, J.J. Beaman, J.W. Barlow and D.L. Boarell, *JOM*, Vo. 42, (4) (1990), 8.
6. A. Lohner, *Metal Powder Report*, Vol. 52, (2) (1997), 24.
7. M.K. Agarwala, D.L. Bourell, B. Wu, and J.J. Beaman, *Proc. Solid Freeform Fabr. Symp*, Ed. H.L. Marcus, J.J. Beaman, J.W. Barlow, D.L. Bourell and R.H. Crawford, The University of Texas, Austin, Texas, 9-11 Aug. 1993, 193.
8. C.R. Dechard, "*Part generation by layerwise selective sintering*", *MS Thesis*, University of Texas, Austin, Texas, 1986.
9. N. K. Vail, J. W. Barlow and H. L. Marcus, *Proc. Solid Freeform Fabr. Symp*, Ed. H.L. Marcus, J.J. Beaman, J.W. Barlow, D.L. Bourell and R.H. Crawford, The University of Texas, Austin, Texas, 9-11 Aug. 1993, 204.
10. H. Naber and F. Breitinger, Anwendungsmöglichkeiten der Rapid Prototyping Technologien für Druckguβwerkzeuge, 17, *Aalener Gieβereisymposium*, 1996.
11. F. Klocke, T. Celiker, Y.A. Song, *RP J.*, Vol. 1, (3) (1995), 32.
12. Y.P. Kathuria, *Surface Coatings Techn.*, Vols. 116-119 (1999), 643.
13. T. Fuesting, L. Brown, S. Das, N. Harlan, G. Lee, J. J. Beaman, J.W. Barlow, D.L. Bourell and K. Sargent, *Proc. Solid Freeform Fabr. Symp.*, Ed. D. Bourell, J.J. Beaman, H.L. Marcus, R.H. Crawford and J.W. Barlow, The University of Texas, Austin, Texas, 12-14 Aug. 1996, 39.
14. M. Shellabear, J. Reichle, and H.L. Langer, *Proc. of the 11th Intern. Symp. for Electromachining*, Ed. J.P.S. Van Griethuyen and D. Kiritsis, Lausanne, Switzerland, 17-21 April, 1995, 681.
15. U. Lakshminarayan, and H.L. Marcus, *Mater. Manuf. Proc.*, Vol. 9, No. 5 (1994), 932.
16. A. Coremans, M. Kauf and P. Hoffmann, *Proc. 5th Eur. Confer. on RP. and Manuf.*, Ed. P.M. Dickens, Helsinki, Finland, 4-6 June, 1996, 195.
17. D. Girouard, *Proc. 4th Inter. Confer. on RP*, Ed. P.C. Richard, A.J. Lightman and J.A. Schenk, Dayton, Ohio, USA, 14-17 June, 1993, 61.
18. A. Coremanx, M. Kauf and P. Hoffmann, *Proc. 5th Eur. Confer. on RP. and Manuf.*, Ed. P.M. Dickens, Helsinki, Finland, 4-6 June, 1996, 245.
19. T. Fuesting, L. Brown, S. Das, N. Harlan, G.Lee, J.J. Beaman, D.L. Bourell and K. Sargent, *Proc. Solid Freeform Fabr. Symp*, Ed. H.L. Marcus, J.J. Beaman, J.W. Barlow, D.L. Bourell and R.H. Crawford, The University of Texas, Austin, Texas, 9-11 Aug. 1993, 42.
20. A. Radon, J. Frydrych and S. Stolarz, *Mater. Sci. Monographs*, Ed. M. M. Ristic, Pul. Elsevier Sci., 1979.
21. C. Wilkening, *Proc. 2nd Intern. Confer. on Development. in RP. and Tooling*, Ed. G Bennett, Buckinghamshire College, UK, 18, 19 Nov. 1996, 153.
22. J.D. Hu, Z.X. Guo, Q.F. Guan and P. Shen, *Sci. Sint.*, Vol. 29, (3) (1997), 195.
23. X. Yang, M. Zhong, T. Zheng and N. Zhang, ICALEO, 1992, 309.
24. R.M. German, *Liquid Phase Sintering*, Plenum Press, New York, 1985.
25. R.M. German and B.H. Rabin, *Powder Met.*, Vol. 28 (1985), 7.
26. B. Van der Schueren, J.P. Kruth, *RP. J.*, Vol.1, (3) (1995), 23.

27. H. Riegger, J. A. Pask, and H. E. Exner, *Sint. Proc.*, Ed. G. C. Kuczynski, Plenum Press, New York, 1980, 219.
28. B. E. Magee and J. Lund, *Z. Meatllk*, Vol. 67 (1976), 596.
29. D.L. Bourell, H.L. Marcus, J.W. Barlow and J.J. Beaman, *Inter. J. Powder Metall.*, Vol. 28, (4) (1992), 369.
30. J. Frenkel, *J. Phys. USSR*, Vol. 9 (1945), 385.
31. A. Belhadjhamida and R. M. German, *Adv. in Powder Metall. Particulate Mater.*, Vol. 3 (1993), 85.
32. F.B. Swinkels and M.F. Ashby, *Acta Metall.*, Vol. 29 (1981), 259.
33. G.W. Scherer, *J. Amer. Ceram. Soc.*, Vol. 60, (5-6) (1977), 236.
34. G.W. Scherer, *J. Amer. Ceram. Soc.*, Vol. 69, (9) (1986), 206.
35. M.M. Sun, J.C. Nelson, J.J. Beaman and J.W. Barlow, *Proc. Solid Freeform Fabr. Symp.*, Ed. J.J. Beaman, H.L. Marcus, J.W. Barlow, D.L. Bourell and R.H. Crawford, The University of Texas, Austin, Texas, 11-14 Aug., 1991, 46.
36. R.M. German, *Intern. J. Powder Metall.*, Vol. 26 (1990), 35.
37. P. Haase, *Selective Laser Sintering of Metal Powders, MS Thesis*, The University of Texas at Austin, TX, 1989.
38. J.A. Manriquez-Frayre, and D.L. Bourell, *Proc. Solid Freeform Fabr. Symp.*, The University of Texas, Austin, Texas, 6-8 Aug. 1990, 99.
39. J.A. Manriquez-Frayre, and D.L. Bourell, *Proc. of Solid Freeform Fabr. Symp.*, Ed. J.J. Beaman, H.L. Marcus, D.L. Bourell and J.W. Barlow, The University of Texas, Austin, Texas, 11-14 Aug., 1991, 236.
40. D. J. Lee and R. M. German, *Intern. J. Powder Techn.*, Vol. 21 (1985), 9.
41. A. P. Savitskii, *Soviet Powder Met. Metal Ceram.*, Vol 26 (1987), 631.
42. S. Farooq, A. Bose, and R. M. German, *Prog. Powder Metall.*, Vol.43 (1987), 65.
43. J. K. Park, S. J. L. Kang, E. Y. Eun, and D. N. Yoon, *Metall. Trans.*, Vol. 20A (1989), 837.
44. J.J. Beaman and C.R. Deckard, *Selecive Laser Sintering with Assisted Power Handling*, US patent No.4,938,816, June 1990.
45 B. Badrinarayan, J.W. Barlow, *Proc. Solid Freeform Fabr. Symp.*, Ed. H.L. Marcus, J.J. Beaman, D.L. Bourell, J.W. Barlow and R.H. Crawford, The University of Texas at Austin, Austin, TX, 8-10 Aug. 1994, 371.
46. J.R. Tobin, B. Badrinarayan, J.W. Barlow, J.J. Beaman, and D.L. Bourell, *Proc. Solid Freeform Fabr. Symp.*, Ed. H.L. Marcus, J.J. Beaman, J.W. Barlow, D.L. Bourell and R.H. Crawford, The University of Texas, Austin, Texas, 9-11 Aug. 1993, 303.
47. B. Badrinarayan, *Study of The Selective Laser Sintering of Metal-Polymer Powders, Ph.D. Dissertation*, The University of Texas at Austin, Dec.1995.
48 J.W. Barlow, J.J. Beaman, D.L. Bourell, R.H. Crawford, H.L. Marcus, J.R. Tobin and N.K. vail, *U.S. Patent Appl.*, 1994.
49. R. Jurand, ed., *Modern Plastics Encyclopedia*, McGraw-Hill: New York, 1989.
50. J.W. Barlow, J.J. Beaman, B. Badrinarayan, *RP J.*, Vol. 2 (1996), 4.
51. D.L.Bourell, H. L. Marcus, J. W. Barlow, J. J. Beaman, C. R. Deckard, *Patent* #4,944,817, issued July 31, 1990.
52. G. Prabhu, and D.L. Bourell, *Proc. Solid Freeform Fabr. Symp.*, Ed. H.L. Marcus, J.J. Beaman, J.W. Barlow, D.L. Bourell and R.H. Crawford, The University of Texas, Austin, Texas, 9-11 Aug. 1993, 317.
53. D.E. Bunnell, D.L. Bourell, H.L. Marcus, *Proc. Solid Freeform Fabr. Symp.*, Ed. H.L. Marcus, J.J. Beaman, D.L. Bourell, J.W. Barlow and R.H. Crawford, The University of Texas at Austin, Austin, TX, 8-10 Aug. 1994, 379.
54. M.K. Agarwala, D.L. Bourell, J.J. Beaman, B. Wu, and H.L. Marcus, *Proc. of the 123rd TMS-AIME Annual Meeting*, Ed.,Gary Warren, San Francisco CA (1994), 833.
55. O. Nyrhila, *US Patent* 5,061,439, (1991).

56. M. Sindel, T. Pintat, M. Greul, O. Nyrhila, and C. Wilkening, *Proc. Solid Freeform Fabr. Symp.*, Ed. H.L. Marcus, J.J. Beaman, D.L. Bourell, J.W. Barlow and R.H. Crawford, The University of Texas at Austin, Austin, TX, 8-10 Aug. 1994, 94.
57. *Metal Handbook*, 9th Edition, Amer. Soc. For Metals, Metals Park, Ohio, Vol. 2 (1979), 777.
58. Z. Guo, J. Hu and Z. Zhou, *J. Mater. Sci.*, Vol. 34 (1999), 5403.
59. *Metal Handbook*, Ninth Edition, Amer. Soc. For Metals, Metals Park, Ohio, Vol. 2 (1979), 470.
60. Data Sheet, *Electrolux Rapid Development*, 1998.
61. *Metal Handbook*, 9th Edition, Amer. Soc. For Metals, Metals Park, Ohio, Vol. 2 (1979), 421.
62. G. Zong, Y. Wu, H. L. Marcus, D. L. Bourell, and J. J. Beaman, *Proc. of the symp. on Proc. and Fabr. of Adv. Mater. for High Tem. Appl. -II*, (V. A. Ravi and T. S. Srivatsan, Eds), Chicago IL, 1-5 Nov., 1992, The Metals, Minerals and Mining Society, 503.
63. A. Manthiram, F. Chi, L.F. Johnson, B.R. Birminnham and H.L. Marcus, *Proc. Solid Freeform Fabr. Symp.*, Ed. H.L. Marcus, J.J. Beaman, D.L. Bourell, J.W. Barlow and R.H. Crawford, The University of Texas at Austin, Austin, TX, 8-10 Aug. 1994, 4.
64. A. Manthiram, D.L. Bourell, and H.L. Marcus, *JOM*, Vol. 45, (11) (1993b), 66.
65. R.D. Cottel, B. Birmingham, Z. Elizer and H. L. Marcus, *Prof. of Solid Freeform Fabr. Symp.*, The University of Texas, Austin, Texas, 7-9 Aug.1995, 369.
66. K. McAlea, R. Booth, P. Forderhase, U. Lakshminarayan, *27th Intern. SAMPE Techn. Confer.*, Oct. 9-12, 1995, 949.
67. U. Hejmadi and K. McAlea, *Proc. Solid Freeform Fabr. Symp.*, Ed. D. Bourell, J.J. Beaman, H.L. Marcus, R.H. Crawford and J.W. Barlow, The University of Texas, Austin, Texas, 12-14 Aug. 1996, 97.
68. U. Lakshminarayan, K McAlea, D. Girouard and R. Booth, *Adv. In Powder Metall. Particulate Mater.*, Metal Powder Industries Federation, Princeton, NJ, USA, Vol. 3 (1995), 13/77.
69. K. McAlea, P. Forderhase, U. Hejmadi, C. Nelson, *Proc. of 7th Intern. Confer. on RP–1997*, San Francisco, California, USA , March-April, 1997, 23.
70. M. Hartley, A. Venus, and S.J.V.D. Crommert, *Initial Experience of Using DTM's RapidTool TM Metal Process*, 237.
71. A.D. Venus, S.J.V.D. Crommert, *Proc. of the 5th Europ. Confer. on RP and Manufacturing*, Ed. P.M. Dickens, Helsinki, Finland, 4-6 June, 1996, 171.
72. A.D. Venus, S.J.V.D. Crommert, *RP and Manuf. '96*, Hyatt Regency, Dearborn, Michigan, USA, April 1996, Vol. 2.
73. J.B. Dyn, *Injection Molds and Molding, A Practical Manual*, 2nd Edition, Van Nostrand Reinhold, New York, (1987), 159.
74. *Metal Handbook*, Ninth Edition, Amer. Soc. For Metals, Metals Park, Ohio, Vol. 3 (1979), 5.
75. Ibid, 18.
76. W. Konig, T. Celiker and Y.A. Song, *Proc of the 3rd Eur. Confer. on RP.*, 6-7 July 1994, Nottingham, 245.
77. S. Das, M. Wohlet, J.J. Beman, D.L. Bourell, *Proc. Confer. on Adv. in Powder Metall. Particulate Mater.s*, Chicago, IL, USA, 6-7, Vol. 3 (1997), 67.
78. J. Wright, R. Knight, J.J. Beaman, D. Freitag, *Prototyping Functional Metal Parts by the SLS/HIP Process*, Lockhea Martin Vought System, 149.
79. J. Beaman, J.W. Barlow, D.L. Bourell, R.H. Crawford, H.L. Marcus and K.P. McAlea, *Solid Freeform Fabrication*, Kluwer Academic Publishers, Norwell, MA, 1997.
80. D.T. Pham, S.S. Dimov and F. Lacan, *IMechE*, Vol. 214, B (2000), 107.
81. S. Das, J.J. Beaman, M. Wohlert and D.L. Bourell, *RP. J.*, Vol. 4, (3) (1998), 112.
82. S. Das, M. Wohlert, J.J. Beaman, and D.L. Bourell, *JOM*, Vol. 50, (12) (1998), 17.

83. S. Das, M. Wohlert, J. J. Beaman and D. L. Bourell, *Mater. Des.*, Vol. 20, (2) (1999), 115.
84. S.T. Mileiko, *Metal and Ceramic Based Composites*, Composites Materials Series, Elsevier Sci., The Netherlands, 12 (1997), 547.
85. T. Krantz, *Intern. J. Powdr Metall.*, Vol. 5 (1969), 35.
86. L. Styger, *IEE Review*, Vol. 41, No. 1 (1995), 38.
87. S. Banerjee, R. Oberacker and C.G. Goetzel, *Intern. J. Poder Metall. Tech.*, Vol. 20 (1984), 325.
88. H.V. Atkinson and B.A. Rickinson, *Hot Isostatic Processing*, Publ. Adam HilgerImprint, IOP Publ. Ltd., (1991).
89. T. Wohlers, *RPA/SME Rapid Prototyping Newsletter*, Vol. 1, (2) (1995).
90. J. Wright, R. Knight, J. Beaman and D.W. Freitag, *Proc. Solid Freeform Fabr. Symp.*, Ed. H.L. Marcus, J.J. Beaman, J.W. Barlow, D.L. Bourell and R.H. Crawford ,The University of Texas, Austin, Texas, 3-5 Aug. 1992, 149.
91. F.P. Knudsen, *J. Amer. Ceram. Soc.*, Vol. 42, (8) (1959), 376.

Chapter 6

METAL-BASED SYSTEM VIA LASER MELTING

6.1 Selective laser melting process

The advantage of SLS is that it uses only low power laser. However, since the powder is not totally melted during laser scanning, the SLS-processed parts are not fully dense and hence, have relatively low strength. To overcome this disadvantage of traditional SLS, selective laser melting (SLM) and selective laser cladding (SLC) processes have been developed that enable full melting of the powder. Basically, the SLM process is the same as SLS except for the much higher laser energy density used. The powder bed is fully or partially melted directly to form metallic bonding [1]. The selective laser cladding (SLC) process is based on the surface cladding treatment.

6.1.1 Selective laser cladding

SLC has been developed over the past decades. It is originally developed for surface treatment, which is to modify the surface properties of the part. In this process, an alloy layer is fused onto the surface of a substrate to create new surface with a metallurgical bonding and with low dilution of substrate [2, 3]. In recent years, laser cladding process has been further investigated to produce 3D metallic parts [4, 5]. The unique advantage of this technique is one-step manufacture of metallurgical uniform 3D objects.

6.1.1.1 Rapid prototyping using selective laser cladding process

Fig. 6.1 shows the principle of SLC. The SLC system consists of the laser, metal powder delivery system, and computer controller. The focused laser beam (and/or XY table) is moved along the contours of the part.

In the SLC process, the powder particles are blown directly into the laser-induced molten pool to form a cladding line. As the laser beam moves away from the pool, the molten material is solidified forming fully dense structure. In the SLC process, the laser beam power, scan speed, and powder feeding rate are key factors in controlling the quality of the SLC part [6]. For a successful cladding process, it is important to melt the powders and homogenize the melt pool. Usually, the high temperature gradient within the melt pool causes intense convection by the Marangoni effect [7]. The homogenization becomes extremely important when a mixture of powders with different compositions is used. This is dependent on the temperature, heat flow, laser beam spot and other parameters. Homogenization is sometimes described by surface tension number *S*, which can be represented as [8]:

$$S = \frac{(\mathrm{d}\gamma / \mathrm{d}T)qd}{\mu Vk} \tag{6.1}$$

where $\mathrm{d}\gamma/\mathrm{d}T$ is the temperature derivative of the surface tension, q, the net heat flow from a laser beam, d, the diameter of the laser beam, μ, the viscosity, V, the scan speed of the laser beam and k, the thermal conductivity.

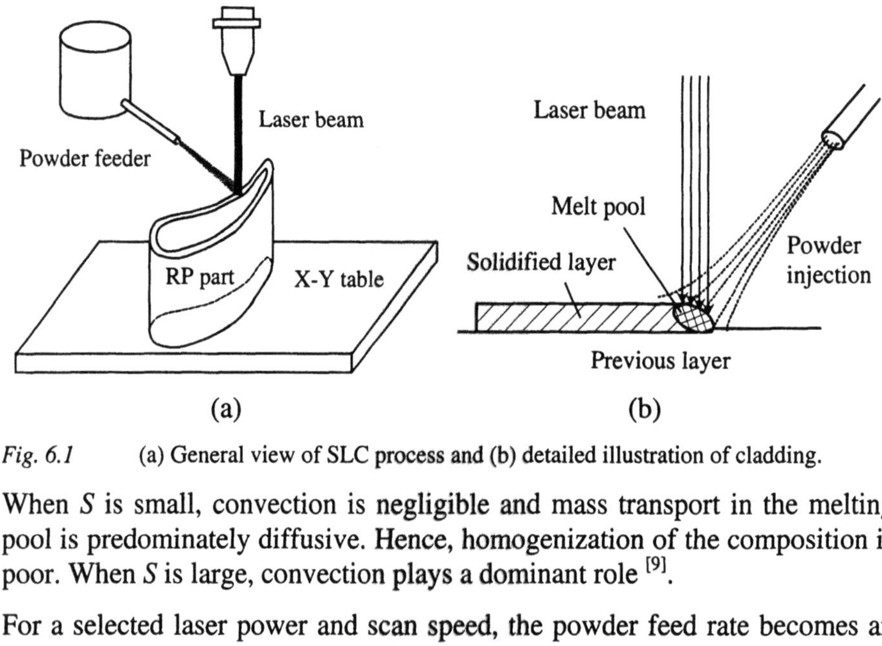

Fig. 6.1 (a) General view of SLC process and (b) detailed illustration of cladding.

When *S* is small, convection is negligible and mass transport in the melting pool is predominately diffusive. Hence, homogenization of the composition is poor. When *S* is large, convection plays a dominant role [9].

For a selected laser power and scan speed, the powder feed rate becomes an important factor. The amount of powder injected to the melting pool is determined by the rate of powder flow. At a low powder feed rate, the clad height and contact angle are low and the melting pool is deep (Fig. 6.2 (a)). When the powder feed rate increases, the laser energy that can be absorbed by the powder per unit volume is decreased. The height and contact angle increases, as shown schematically in Fig. 6.2 (b) [10].

In the SLC process, the width of a cladding tract is determined mainly by the laser beam size, whereas the height is determined by the laser power density, scan speed and powder feed rate. The horizontal overlapping pitch is a crucial factor to achieve a better surface finish and minimized porosity defects of the final object [11].

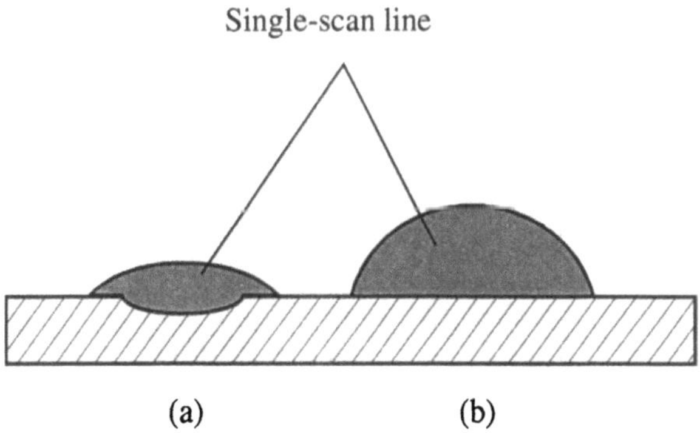

Fig. 6.2 Influence of powder feed rate. (a) low powder feed rate and (b) high powder feed rate.

6.1.1.2 Design of nozzle

In some laser cladding process, the powder is injected laterally into the interaction region through an injection tube (Fig. 6.1). The main disadvantage of the lateral powder feeding system is that the powder is not axially injected into the melt surface in the plane of the laser beam and the scan direction. To achieve such deposition of the powder at the melt surface, coaxial nozzles have been developed [12-14].

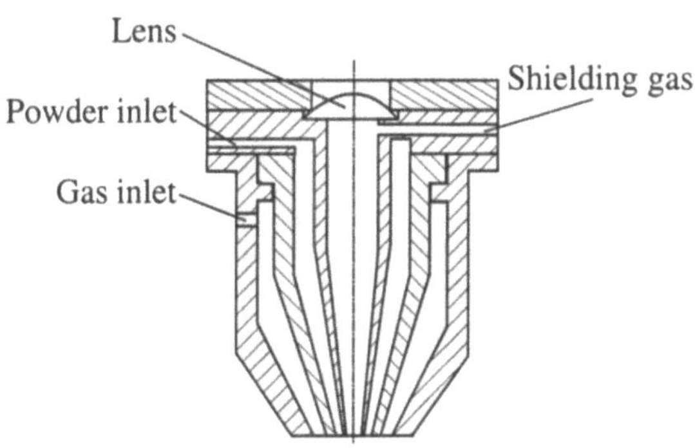

Fig. 6.3 Schematic illustration of a coaxial nozzle.

Fig. 6.3 schematically shows a coaxial nozzle. The laser beam passes though the center of the nozzle. The lens is protected by a shielding gas. The powder particles are injected by inert gas flowing at low speed through the middle cone and another protective gas flowing through the outer cone. The focused laser beam is directed onto the substrate surface with simultaneous coaxial feeding of the metallic cladding powder to produce a fine clad track.

Fig. 6.4 shows the building of a thin-wall part using a 200-watt sealed CO_2 laser and a coaxial nozzle developed at the Gintic Institute of Manufacturing Technology [11]. With optimized parameters and proper set-up, a thin beam cladding process can be obtained and used to build up 3D objects. A clad track of about 300 μm wide has been achieved under a focused laser beam. The build-up rate is about 1.7 cm^3/h.

Fig. 6.4 Rapid prototyping using a coaxial nozzle [11].

Many different material systems [11, 16-19] have been developed and used in the SLC process.

The advantage of using the SLC process is its flexibility in alloying using different powder feeders. However, due to the nature of the powder delivery system, it is difficult to build a part with curvature or complicated geometry, such as the part shown in Fig. 6.5, which consists of two plates connected by a sphere ball on which a cylinder is attached. Since there is no support system, only parts with simple and straight features can be fabricated.

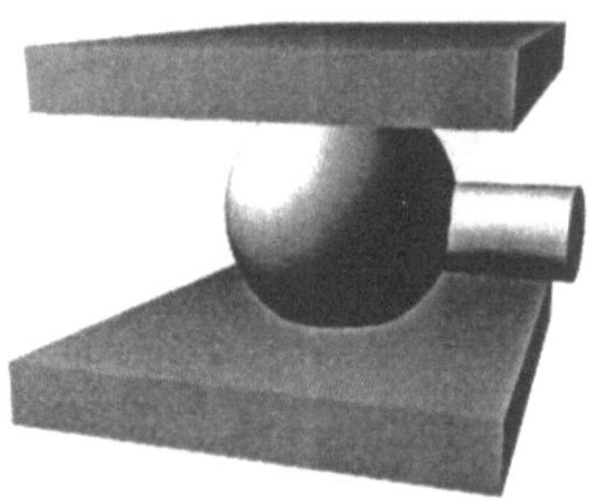

Fig. 6.5 Curved geometry.

6.1.2 Selective laser melting

6.1.2.1 Rapid prototyping using selective laser melting process

Selective laser melting (SLM) is very similar to SLS as both are based on the same concept. The only difference is partial melting of the powder in SLS and full melting in SLM. As in SLS, in SLM, a layer of powder is pre-deposited onto the previous layer or substrate and compacted by a roller. The high power density laser fully melts the pre-deposited powder layer according to a specific computer-generated pattern. Fig. 6.6 schematically illustrates the SLM process where a portion of the deposited powder particles is melted to form a melting pool.

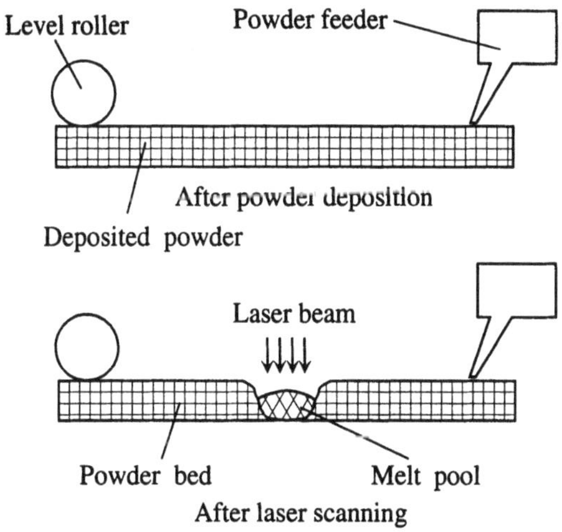

Fig. 6.6 Schematic view of selective laser melting process.

Since the powder is fully melted in the SLM process, protection of the SLM-processed part from oxidation is essential. In addition, due to full melting, there can be large shrinkage after solidification. Distortion, and even crack, and inaccuracy due to thermal dilatation become a major issue. To overcome the distortion and crack, the base metals and/or even the whole chamber (in which the powder is processed) are often preheated to a certain temperature [20]. Another method is to anchor by sintering the first layer and any overhangs to the substrate.

6.1.3 Process control

6.1.3.1 Laser power and energy density

Density of laser power is one of crucial factor. However, it does not directly affect the properties of fabricated parts as the energy of a laser beam is given by its energy density:

$$E_S = \frac{P}{V\delta} \qquad (\mathrm{J/cm^2})$$

Experimental results have shown that although the same energy density E_S can be obtained by varying different combinations of laser power P, scan speed V and laser beam diameter δ, very different structures and properties can be obtained. As shown in Fig. 6.7 where CuSn 89/11 is used, melting (sintering) depth increases with increase in laser energy density due to increase in laser power or scan speed. At the same laser energy density, for example $E_S = 1.4$ or $E_S = 1.8$, the laser power has much stronger effect on the depth of melting. The size of the laser beam has almost no effect on the depth of melting; in fact, the depth of the melting even decreases slightly as laser beam diameter decreases.

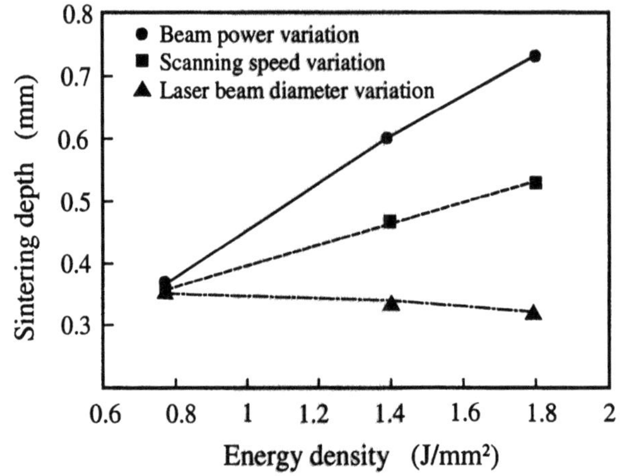

Fig. 6.7 Influence of energy density on sintering depth [21].

The density of the processed parts always increases as the laser energy density increases (Fig. 6.8.). At the same laser energy density, the scan speed shows the strongest influence on the density, while the laser beam power shows the least effect. This is because the laser melt lines are less broken with reduced scan speed and more closely connected with decreasing beam size [21].

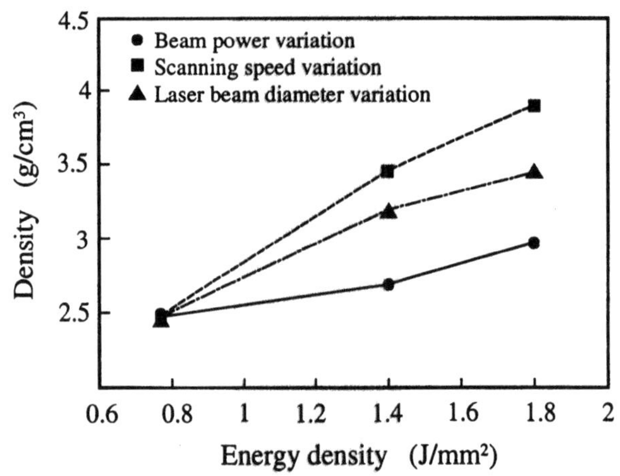

Fig. 6.8 Influence of energy density on the density of processed part [21].

6.1.3.2 Scan speed

When the scan speed is increased, the laser energy density decreases, leading to a reduction in the width of the scanned line. Fig. 6.9 reveals the influence of scan speed on the morphology of single-scanned lines of gas-atomized M2 high-speed steel powder. At a low scan speed of 5 mm/s, both scan lines at laser powers of 50 and 150W are continuous [22]. When the scan speed is increased to 20 mm/s, neither of them shows continuous line. On the other hand, discontinuous line, in the form of balling, has been found in the M2 high-speed steel powder if the scan speed is equal or higher than 10 mm/s at 150 W [22].

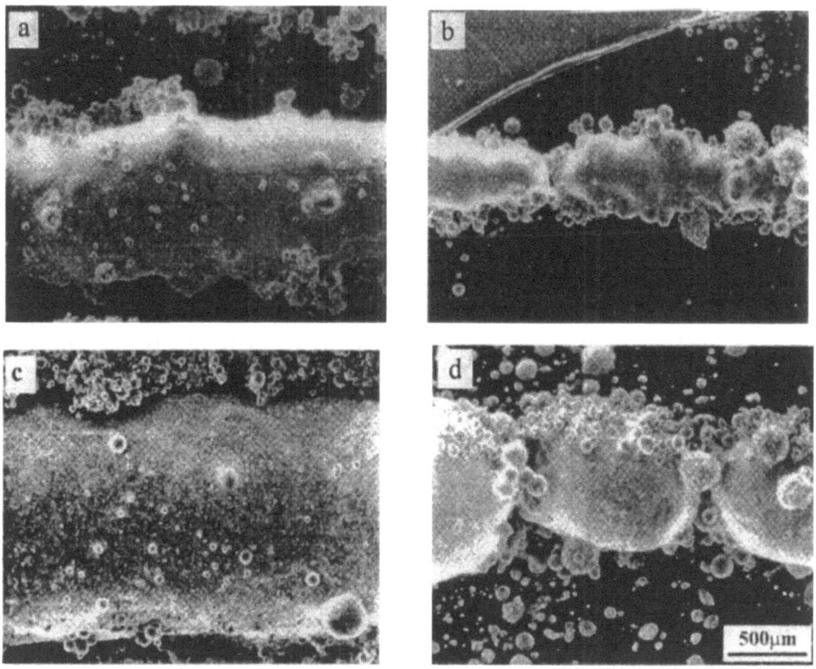

Fig. 6.9 Single-scanned lines showing the influence of scan speed and laser power: (a) 5mm/s, 50W, (b) 20mm/s, 50W, (c) 5mm/s, 150W and (d) 20mm/s, 150W [22].

Based on Eq. (5.1), the laser energy density is calculated with a laser beam diameter of 0.5 mm and listed in Table 6.1. From Fig. 6.9 and Table 6.1, it can be seen that although the laser energy densities for processes No.1 and No.5 are very different, the molten volumes for both processes are almost the same. In addition, although the laser energy density of process No.5 is three times higher than that of process No.2, there is no indication of reduction in the formation of balls in process No.5. This observation indicates that the scan speed is an important factor in determining balling, or the formation of balls.

Table 6.1 Laser energy density

No.	Operation parameter	Laser energy density (J/cm^2)
1	2 mm/s, 50 W	5000
2	20 mm/s, 50 W	500
3	2 mm/s, 150 W	15000
4	10 mm/s, 150 W	3000
5	20 mm/s, 150 W	1500

The balling effect is associated to instability of the molten track. During laser melting, the liquid track tends to break up into a row of spheres to reduce its surface energy [23, 24]. Analysis by Rayleigh [25] shows that a water cylinder is unstable in front of a sinusoidal wave fluctuation if the wavelength $\lambda > \pi D$, where D is the initial diameter of the unperturbed cylinder. With increase in the initial diameter D, the time required to break up the cylinder increases [26]. In laser melting, a steep thermal gradient is developed between the center and edge of the molten track. Since surface tension is a function of temperature, the presence of a temperature gradient causes a corresponding variation in the surface tension between the center and the edge of the molten track. The gradient of the surface tension will induce a Marangoni flow from a region of low surface tension to a region of high surface tension. Therefore a force will be induced that acts on the molten track [22]. Finally if the force is too high, the continuous molten tract may be broken up to form balls. When the molten track is broken up into agglomerates, the volume of the molten track is identical as the agglomerates so that the diameter of the agglomerates is proportional to that of the molten track [27]. Obviously from Fig. 6.9, the diameter of the molten track increases with increase in the laser power and decrease in the scan speed. Therefore, by increasing the laser power or decreasing the scan speed, the molten track will not break up easily into the agglomerates so that continuous lines can be obtained [27]. Fig. 6.10 shows the influence of laser scan speed and laser power on the variation of the agglomerate diameter [27]. The diameter only increases slightly with decreasing scan speed until a critical speed. After that, the agglomerate increases rapidly. The diameter of the agglomerate has been experimentally demonstrated to be a linear function only of the laser power.

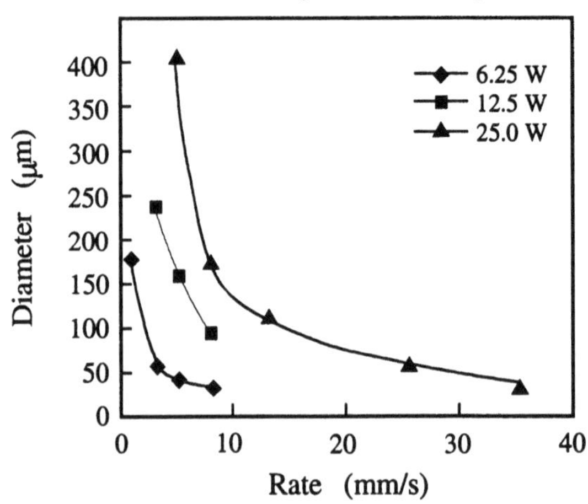

Fig. 6.10 Effect of laser scan speed and laser power on the variation of the agglomerate diameter [27].

Compared with laser power, the effect of scan speed on balling is more dominant. A low scan speed tends to keep the instantaneous liquid track of $\lambda < \pi D$, and hence a smooth surface can be obtained [27]. Therefore, a low scan speed is preferred. Furthermore, the tendency for the formation of balls is also closely associated with the oxygen content in the powder. The formation of oxides may increase the surface tension temperature coefficient across the surface and change its sign from a negative to a positive value [22, 28].

Niu *et al.* [29] has developed a map by which correct combinations of laser powers and scan speeds can be selected. Fig. 6.11 shows the variation of porosity in the laser-formed surface when M2 high-speed steel powder is processed using a CO_2 laser of 0.5-mm beam diameter with different laser power and scan speed. At extremely low laser power and scan speed, big open pores are formed. With increasing scan speed, the big open pores change to small closed pores. When the laser power is above 40 W, a fully dense part can be fabricated with a low-scan speed. If a high scan speed is required, the laser power has to be increased much more to ensure low porosity.

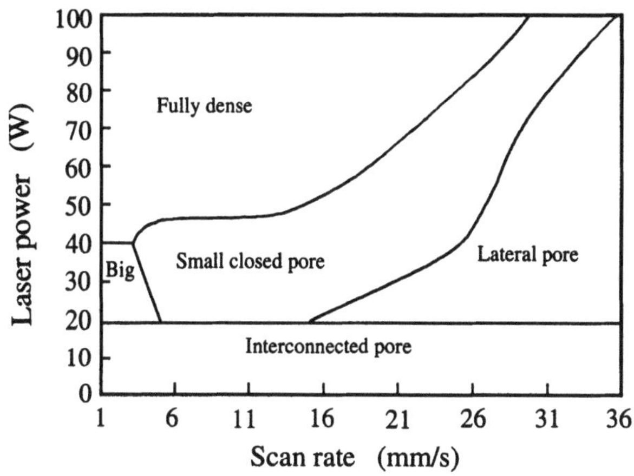

Fig. 6.11 Variation of porosity with laser power and scan speed [29].

6.1.3.3 Scan pitch

The scan pitch must be smaller than the track width (Fig. 6.12) to ensure good metallic bonding between neighboring tracks and minimum porosity. Since the width of a track is not only determined by the laser beam size but also by the laser energy density, scan speed, and powder property, scan pitch cannot be simply defined by the laser beam size.

Fig. 6.13. schematically illustrates the influence of laser energy density and scan speed on the width of a track. At low laser energy density and/or fast scan speed, the width of the track is smaller than the diameter of the laser beam. When high laser energy is used, the width of the track may be wider than that of laser beam size. In general, at least 30% overlap is essential to ensure good structural property [30].

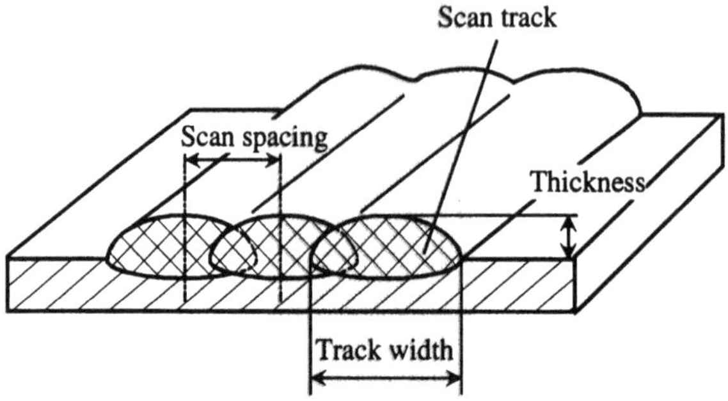

Fig. 6.12 Geometry of scanned track.

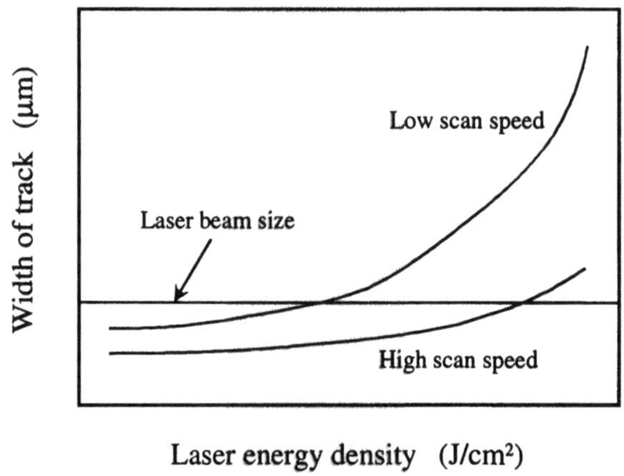

Fig. 6.13 Change of the track width as a function of laser energy density and scan speed.

6.1.3.4 Thickness of track

The depth of laser-melted track can be increased by increasing the laser energy density and/or decreasing the scan speed. Although fabrication time can be shortened, both accuracy and surface finish of the fabricated part are degraded as layer thickness is increased. To obtain good metallic bonding, the track being scanned must re-melt a portion of previous layer. Fig. 6.14 shows two types of scanned layers. Although the layer thickness in both cases is almost the same, there is only a very small overlapped area in case shown in Fig. 6.14 (a) where the creation of the layer is mainly due to melting of the deposited powder. In Fig. 6.14 (b), the thickness of deposited powder layer is much thinner than that in Fig. 6.14 (a). Therefore the creation of the layer is partly from the deposited powder and partly from re-melting of the previously formed layer, giving better bonding.

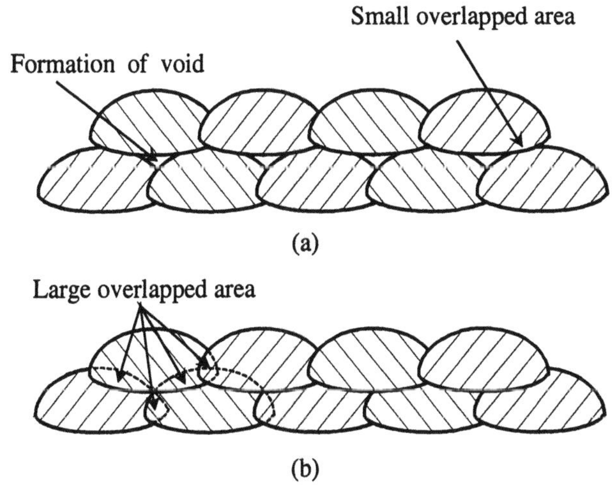

Fig. 6.14 Two types of scanned layers: (a) poor bonding between the two adjacent layers and (b) good bonding between the two adjacent layers.

6.2 Metal powders

6.2.1 Ti system

One successful example of using the cladding process is in the fabrication of titanium aerospace components by Lasform^{SM}. This process was developed more than 20 years ago, and technical feasibility and scale-up were demonstrated by the U.S. Defense Advanced Research Projects Agency [16].

Ti-6Al-4V RP part is formed in a protective atmosphere of argon that contains less than 10 ppm of oxygen. This is critical for Ti due to its high susceptibility to oxygen and nitrogen contamination [17]. The microstructure of the laser-clad Ti-6Al-4V consists of relatively fine prior β grains that are about the same size as those in thin casting. Due to the fast cooling rate used, the integrity of the clad part shows as good as, or even better than, that of cast and wrought products. It is claimed that the SLC processed parts are 100% defect free. Table 6.2 shows mechanical properties of Lasform Ti-6Al-4V,which are equivalent to those of cast and wrought materials.

Table 6.2 Mechanical properties of Lasform Ti-6Al-4V [17]

Property	Value
Ultimate tensile strength	1030 MPa
Yield strength	900 MPa
Elongation	12.3%
Reduction in area	23.5%
Fracture toughness	90 MPa-m$^{1/2}$
Charpy impact strength	19 J
Hardness	36 HRC

Smooth S-N fatigue data in Fig. 6.15 shows much better property than that of cast Ti-6Al-4V [31].

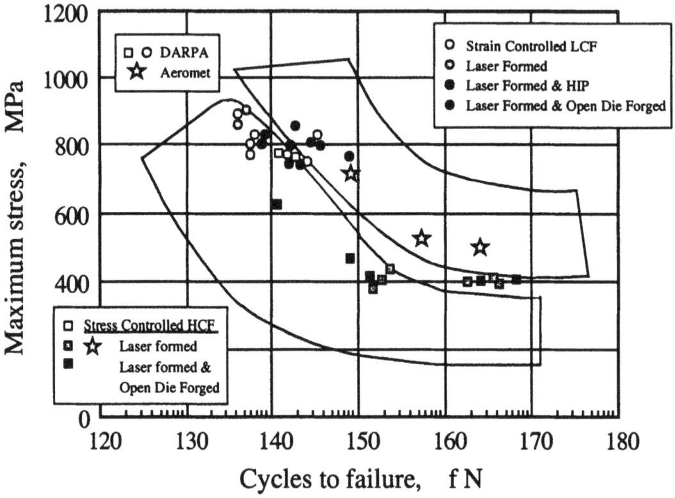

Fig. 6.15. S-N fatigue data of Lasformed^{SM} Ti-6Al-4V [31].

Table 6.3 [18] compares the mechanical properties of wrought material counterparts with three different materials fabricated using a SLC process developed at both Loa Alamos and Sandia National Laboratories. The mechanical properties of these materials are in general better than the wrought material counterparts. The increase in the yield stress is due mainly to fine grain size and cellular microstructure.

Table 6.3 Tensile properties of RP parts processed using DLF and LENS^{TM} process compared to conventionally processed materials [18]

Material	Yield strength MPa	UTS MPa	Elongation %
316 Stainless steel			
RP processed	296	579	41
Wrought annealed	262	572	63
Investment case	269	517	39
Inconel 690			
RP processed	450	666	48.8
Hot rolled rod	372	738	50
Ti-6Al-6V			
RP processed	958	1027	6.2
Wrought bar	827 - 1000	931 - 1069	15 - 20
Cast + anneal	889	1014	10

There are in general two types of defects, namely micro-cracks and micro-voids. Micro-cracks are due to thermal stress. Micro-voids are primarily formed by gas and moisture evaporation during solidification, and lack of fusion between layers of adjacent tracks of the molten pool.

6.2.2 Iron-based system

Iron-based alloys are widely used in die and mold making due to their low cost, high strength, good wear resistance, and relatively high temperature resistance.

Pure Fe melts at 1538°C and boils at 2253°C. However, it is not suitable for structural application and must be alloyed. Basic alloying element is carbon varying from 0.15 wt.% to 1.8wt.% depending on usage of the steels. In addition to carbon alloying element, other types of elements, such as, Cr, Mn, Ni and Si, are also very often used. The melting temperature of alloyed iron changes with the type of alloying addition, as well as with the amount of addition. Some alloying addition does not change the melting temperature of the alloys too much, while others may dramatically affect the melting temperature. For example, addition of Cr does not have much effect on the melting temperature.

CO_2 laser can melt Fe and its alloys earlier since Fe and its alloys can well absorb laser energy. To obtain a homogeneous chemical distribution, prealloyed powder or well-milled powder mixture can be used for SLM process. Fig. 6.16 shows single-line scanned parts using a CO_2 laser operating at power P of 200W, scan speed V of 200mm/min and laser spot size δ of 1mm. Microstructure of SLM processed part of prealloyed Fe15%Cr1.5B powder shows fullly dense structure (Fig. 6.17). Due to the fast solidification rate, fine cellular microstructure is formed. More detailed microstructural evaluation shows needle-like structure between the cellular cells, as shown in Fig. 6.18. The needle-like structure is martensite that is formed because of the fast cooling rate.

Fig. 6.16 Single-line scanned parts.

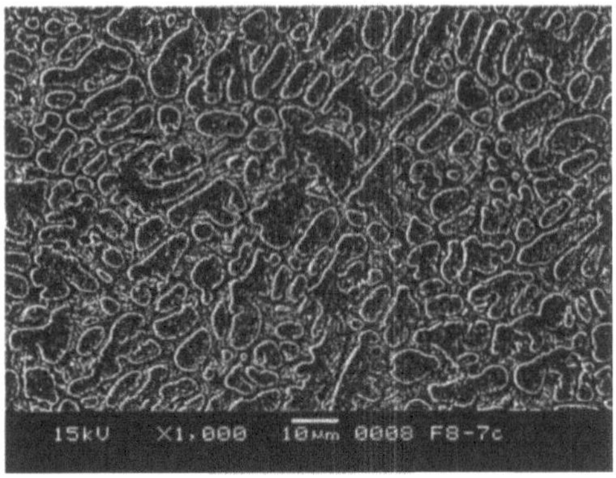

Fig. 6.17 Microstructure of SLM processed Fe15%Cr1.5%B material.

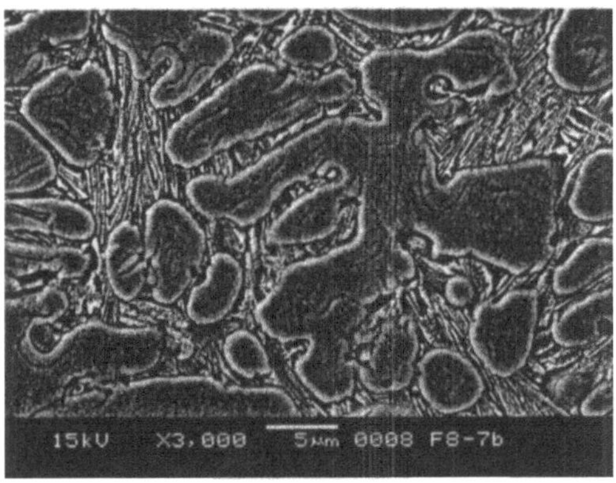

Fig. 6.18 Detailed microstructure of SLM processed Fe15%Cr1.5%B material.

In general, hardness shows an increasing trend with laser scan speed as shown in Fig. 6.19. In this case, the highest hardness of about HRc45 can be obtained at P = 1000W and V = 300mm/min. Under a laser power of 600W, a fully dense structure can be obtained (as shown in Fig. 6.17 and 6.18) so that further increase in laser power may not result in increase in density, while high laser power may increase the heat-affected zone and hence, grain size. Fig. 6.20 schematically shows the effect of laser power. Under a high laser powder, a large amount of heat flows to the already solidified portion, which may lead to coarsening of grains. As predicted by Hall-Petch equation, the yield strength σ of a material is a function of grain size:

$$\sigma = \sigma_0 + kd^{-1/2} \tag{6.2}$$

where σ_0 is frictional stress, k, constant and d, grain size. The increase in grain size implies decrease in dislocation density, which in turn reduces the strength of the material.

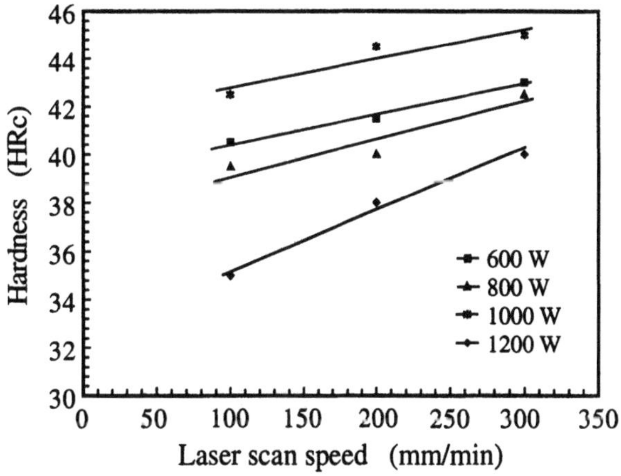

Fig. 6.19 Hardness of SLM processed iron-based parts.

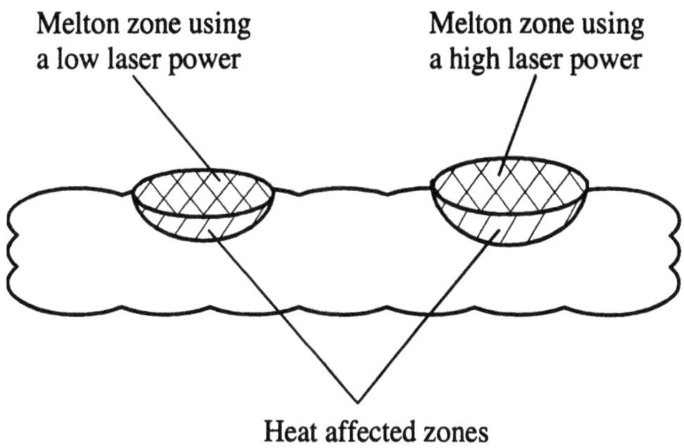

Fig. 6.20 Effect of laser power on the size of the heat-affected zone.

The hardness of the nickel-based material shows the same trend as the iron-based one (Fig. 6.21) with the part processed at a laser power of 800W showing the highest hardness value.

Wear resistance is a crucial consideration if the processed parts are to be used as dies and molds. Fig. 6.22 shows the wear pin-on-disc test result. Two kinds materials are used: conventionally processed P20 alloy bar and SLM-processed pin. The loss in volume of the SLM processed samples is greater than that of the conventionally processed P20 tool steel. The sample using 600W laser power shows the best wear resistance. As the laser power is increased, the volume loss also increases. The increase in volume loss at high laser power may be attributed to the area of heat-affected zone.

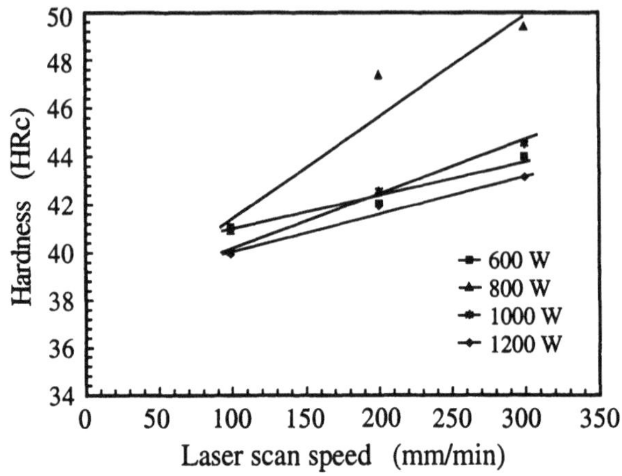

Fig. 6.21 Hardness of selective laser melting processed nickel-based parts.

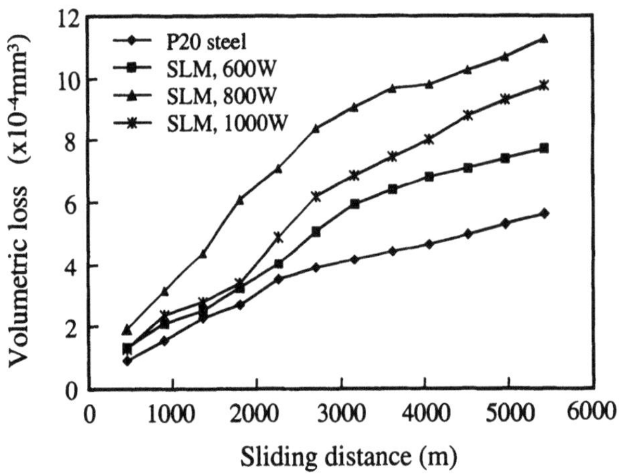

Fig. 6.22 Wear test result of conventional P20 tool steel and SLM processed parts.

Wear result in terms of volume loss per unit sliding distance, as shown in Fig. 6.23, reveals the more visual wear character of SLM-processed iron-base alloy using a laser power of 600W at different laser scan speeds. In general, the wear rate of the SLM processed part is greater than that of the P20 steel. Under the wear condition of 5 kg load, the wear rate of sample laser-scanned at 200mm/min shows as much as 44.2% higher than that of the P20 tool steel. As the laser scan speed is increased, the wear rate is reduced, i.e. the wear resistance is improved. These results confirm the conclusion obtained from Fig. 6.22 that wear rate increases with the increase in laser power. The laser energy density decreases as the laser scan speed increases, which is equivalent to decrease in laser power at constant laser scan speed.

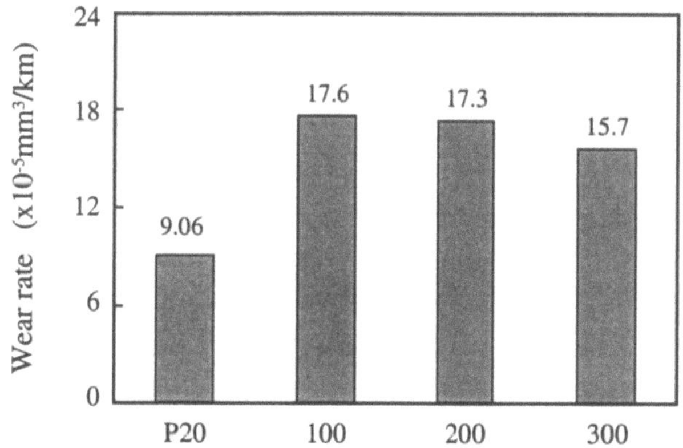

Fig. 6.23 Comparison of wear rate of P20 steel with SLM-processed specimens using different scan rates of 100, 200 and 300 mm/min.

Fig. 6.24 shows the surface morphology of the SLM-processed iron-based alloy specimens and P20 material after wear test. The grinding crack of the SLM-processed specimen is much larger than that of the P20 specimen. The

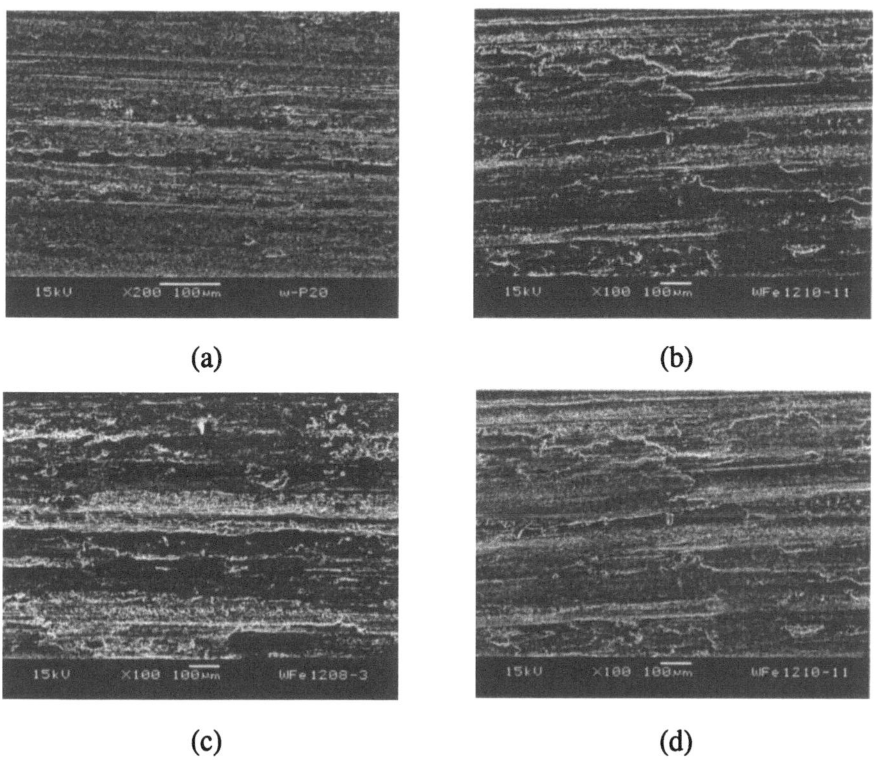

Fig. 6. 24 (a) Surface morphology of wear-tested P20 tool steel and SLM processed specimens at a constant scan speed but different laser power: (b) $P = 600W$, (c) $P = 800W$ and (d) $P = 1000W$.

worn surface shows sliding wear characteristics, which are denoted by repeating plastic deformation tracks. For the specimens prepared using different laser parameters, the surface morphologies are almost the same. No obvious difference in wear mechanism can be found. However, the worn surface of the specimen using 800W laser is slightly greater than the others, which corresponds to the wear rate test. Cavities can be seen in the worn surface from all specimens. These cavities might have resulted from some porosities and impurity. In addition, cracks are found in the vicinity of the cavities

6.2.3 Copper-based system

Rapid tooling is the concept of producing tools through the aid of rapid prototyping. One such application is in electric discharge machining (EDM) [32]. A major cost and time element of EDM is manufacture of the EDM electrode that can account for over 50% of the total machine cost [33]. EDM allows the convenient production of complex shapes in the tool cavity as complex topographies can often be more easily machined on the electrode than inside a cavity. In EDM, there is no direct physical contact between the electrode and the workpiece, and therefore no mechanical stress on the workpiece. This is a very suitable method to shape hard and brittle refractory materials on condition that the electrical resistivity of the material is below 100 $\Omega.cm^2$ [34]. The materials suitable for electrodes should exhibit low wear rates and resistivity, and high thermal conductivity. Traditionally, the electrodes are fabricated from various metallic and nonmetallic materials including copper, tungsten, brass, steel and graphite. The properties of the materials commonly used for EDM electrodes are given in Table 6.4.

Table 6.4 Properties of Cu [26], ZrB_2, TiC, TiB_2 and W [26]

Material	Density (g/cm^3)	Melting point (K)	Electrical resistivity (Ω.cm)	Thermal conductivity (W/m.K)	Thermal expansion coefficient (K^{-1})
ZrB_2	6.09	3300	9.2×10^{-6}	24	5.5×10^{-6}
Cu	8.96	1358	1.69×10^{-6}	401	17×10^{-6}
W	19.3	3410	5.4×10^{-6}	173	4.5×10^{-6}
Mo	10.22	2617	5.7×10^{-6}	138	5.1×10^{-6}
TiC	4.93	3250	215×10^{-6}	17.2	7.4×10^{-6}
TiB_2	4.50	2980	30×10^{-6}	65	4.6×10^{-6}

6.2.3.1 Cu-Ni system

Since it is difficult to laser melt Cu, another element is often added as additive. These additives include Sn, Fe, Ni, etc. Fig. 6.25 shows ball-milled Ni and Cu powder particles. Mechanical milling enables powder particles to be mechanically bonded together, e.g. Cu with Ni particles so that heat can be easily transferred from the Ni to Cu particles when the laser energy is absorbed by the Ni particle that may melt before the Cu particle does.

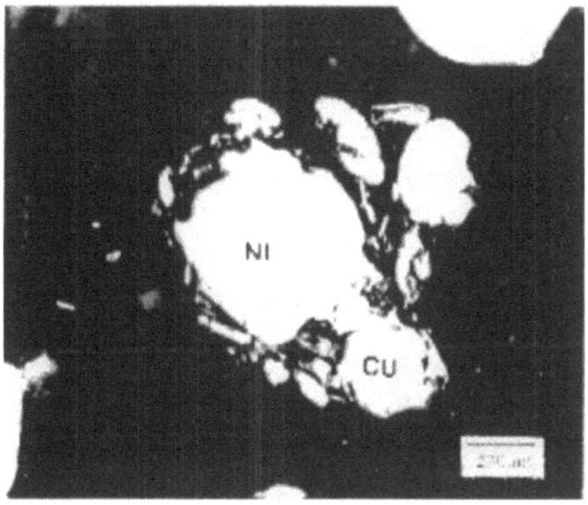

Fig. 6.25 Ball-milled Cu and Ni powder particles.

The Cu-Ni system shows full mixing within the high and low temperatures between the two constituents (Fig. 6.26) without formation of intermediate phases. Hence, it is a good system to be considered. Fig. 6.27 shows the XRD spectra of Cu-Ni system without introduction of Ti and C. Two sets of XRD peaks can be easily distinguished before SLM. Since both Cu and Ni are fcc structure, the two sets of (111), (200) and (220) peaks are close to each other. After SLM, only one set of diffraction peaks can be seen, (while both Cu and Ni diffraction peaks disappear), indicating clearly the formation of Cu-Ni solid solution by SLM.

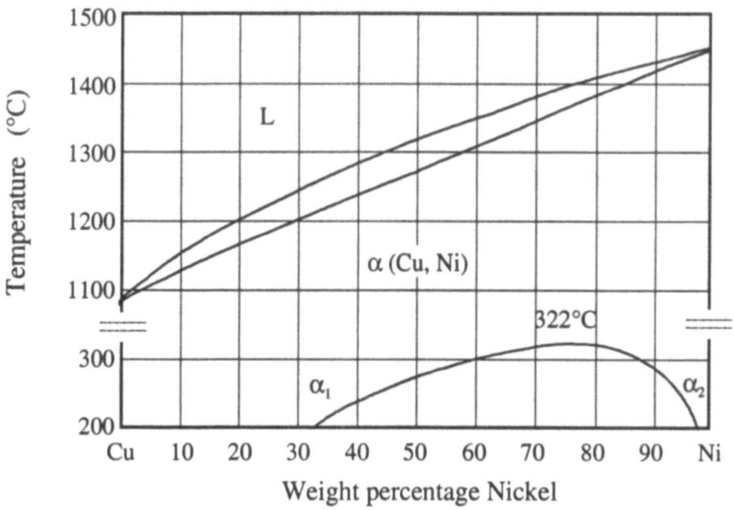

Fig. 6.26. Cu-Ni phase diagram.

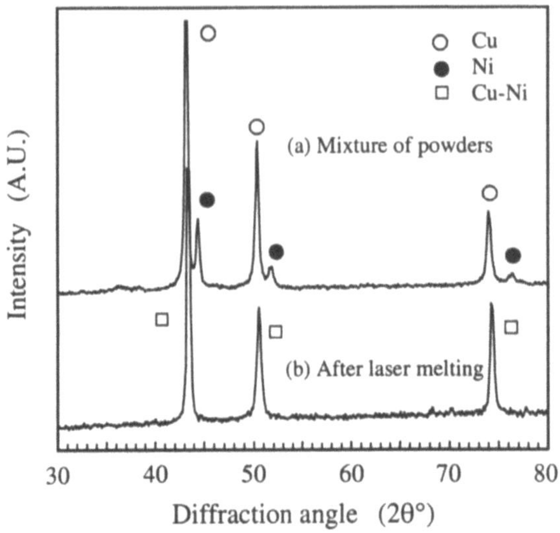

Fig. 6.27 Formation of Cu20wt.%Ni alloy.

Fig. 6.28 (a) reveals the microstructure of a cross section of an SLM processed Cu-Ni line. Two separated α_1 and α_2 phases can be identified. Since Ni content is low, the composition is outside the miscibility gap boundary. The dark α_2 phase shows discontinuity form. No clear grain boundaries can be revealed from the microstructure. When the Ni content is high (Fig. 6. 28 (b)), the grain boundaries can be seen by isolation of α_2 phase along the grain boundaries. This is because a spinodal decomposition brought by compositional fluctuations that take place spontaneously with a decrease in free energy. A few defects in terms of voids can be seen in Cu40wt.%Ni system. This type of voids is common in the structures if moisture of the powders is not well controlled. The vapour from the evaporation of the moisture during melting is sometimes entrapped in the structure.

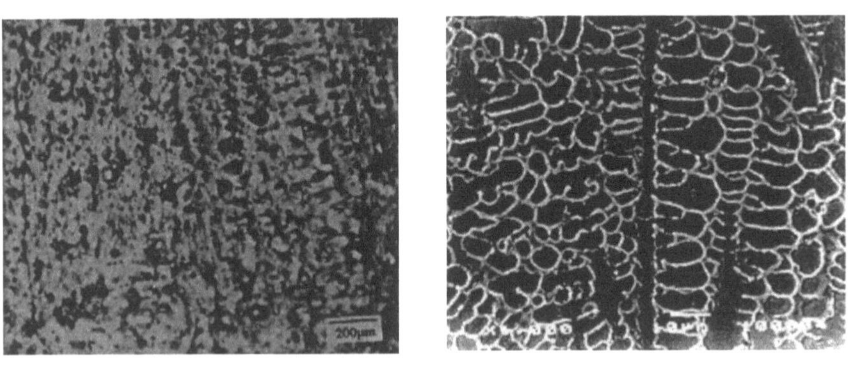

(a) (b)

Fig. 6.28. Microstructure of Cu-Ni alloys: (a) Cu10wt.%Ni and (b) Cu40wt.%Ni.

6.2.3.2 Cu-W system

W is another element often used as EDM electrode material. W has extremely low solubility in the liquid Cu (10^{-5} at.% at 1200°C) [35] and Cu-W is an immiscible system. During the melting of this system, phase separation is expected during solidification. Melting of W is possible if high laser energy density is used. W can also be used as thermal conductor of laser energy to enable melting of Cu. Porosity may be high if W content is too low or too high. The x-ray diffraction pattern of laser-melted Cu10wt.%W is shown in Fig. 6.29. Since there is no solid solution, Cu and W peaks can be observed.

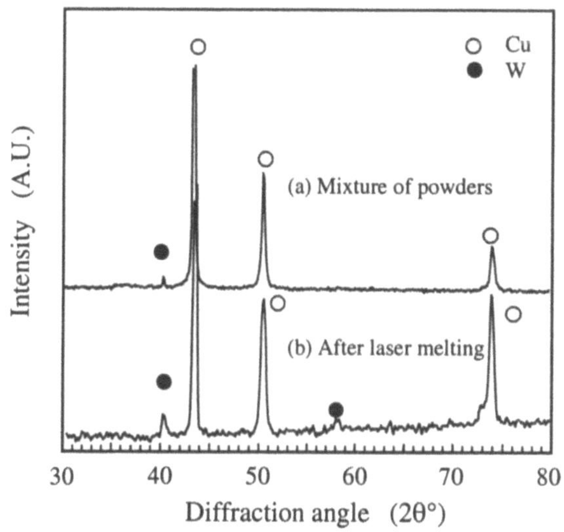

Fig. 6.29 XRD diffraction pattern of Cu10wt.%W.

For the Cu5wt.%W system, when the powder is scanned by the laser beam, the powder consolidates into spheres (Fig. 6.30). The molten metal is completely contained by loose powder rather than fully dense material, and this powder cannot exert tensile traction on the liquid to confine the melt in a layer-wise geometry [23]. The percentage of the high-melting point component W is therefore insufficient for the liquid Cu to wet and bind the unmelted particles together to form a solid line.

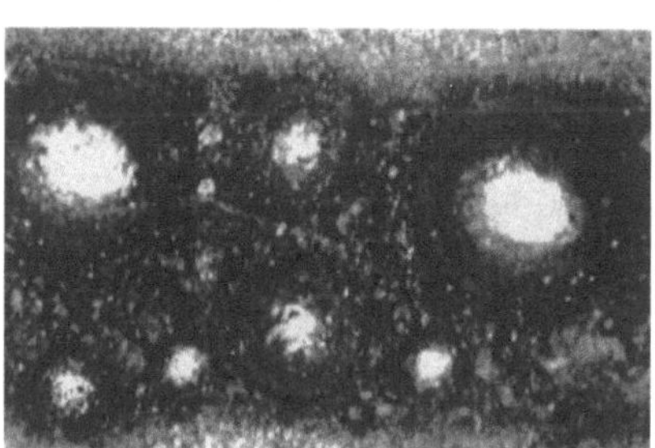

Fig. 6.30 Initial formation of balls.

The cross-section microstructure of the single scanned line of Cu10wt.%W is shown in Fig. 6.31. Large W particles are surrounded by Cu phase. If the W particles agglomerate together, large porosity can be seen as the low-viscosity Cu cannot penetrate into the W agglomerate. The increased amount of melt flow induced by the higher temperature causes the W particles to segregate to the bottom of the single scanned lines due to differences in density between the reinforcement and matrix (ρ_W = 19.3 g/cm^3, ρ_{Cu} = 8.96 g/cm^3), forming a W skeleton, as shown in Fig. 6.32.

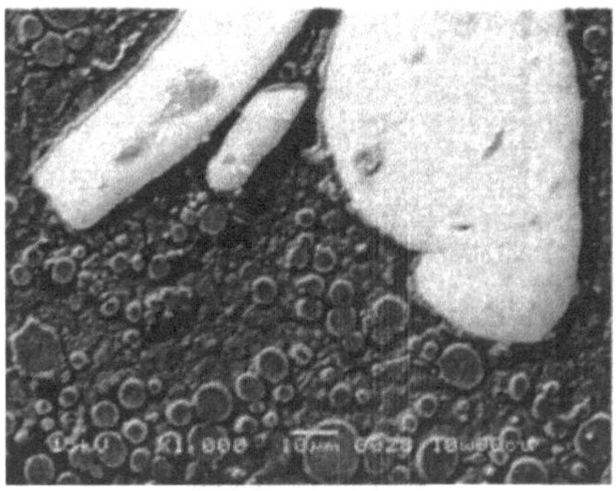

Fig. 6.31 Microstructure of the Cu10wt.%W specimen (*P* = 1000W, *V* = 300mm/min, δ = 3.5mm).

At low percentage of W, the characteristically high porosity of the sample is caused by the high reflectivity of Cu to the laser energy. Thus Cu does not melt easily when irradiated by the laser beam. with The high thermal conductivity of Cu also enhances heat transfer away from the melt zone. This causes the melt to cool rapidly, leading to short rearrangement time and consequently high porosity. Fig. 6.33 compares the reflectivity of some materials as a function of wavelength [36]. The continuous CO_2 laser has a wavelength of 10.6 μm. If it is used on Cu, about 90% of laser energy will be reflected. Although Cu absorbs about 10% of laser energy, the high thermal conductivity may lead to immediate loss of heat.

Fig. 6.32 Microstructure of Cu10wt.%W showing the segregation of W particles (*P* = 1000W, *V* = 250mm/min, δ = 3.5mm).

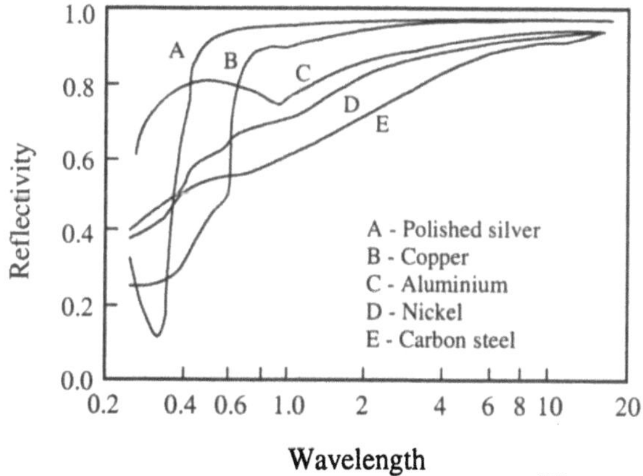

Fig. 6. 33 The relationship between reflectivity and wavelength [36].

Fig. 6.34 (a) shows the microstructure of Cu with 40wt.% of W. Compared with Fig. 6.32, porosity is largely reduced. This can be contributed to the heat transfer from W particles to Cu, which is subsequently melt by the heat obtained from W. From Fig. 6.34, it is clear that the W phase does not melt since the W particles are in rectangular shape. Since most of W particles remain in solid state during laser irradiation due to its high melting point, they act as nucleating centers for Cu to melt. Heterogeneous nucleation requires relatively good wetting between the nucleate and the solid metal. When the percentage of W is higher, the heat absorbed by the W particles is more readily transferred to the Cu particles and caused melting. Fine W particles may have even acted as nucleates for the heterogeneous nucleation of the Cu solid, forming equiaxed structures.

Some area, which is normally on top of a single scanned line, shows less W particles. Microstructure of the specimen, as revealed in Fig. 6.34 (b), shows equiaxed Cu grains of approximately 4 to 10μm. However, when laser power increases, the grain size also increases due to the increase in temperature of the metal bed.

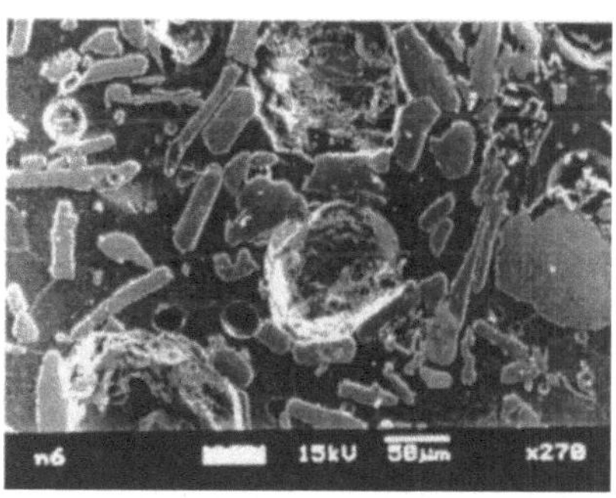

(a)

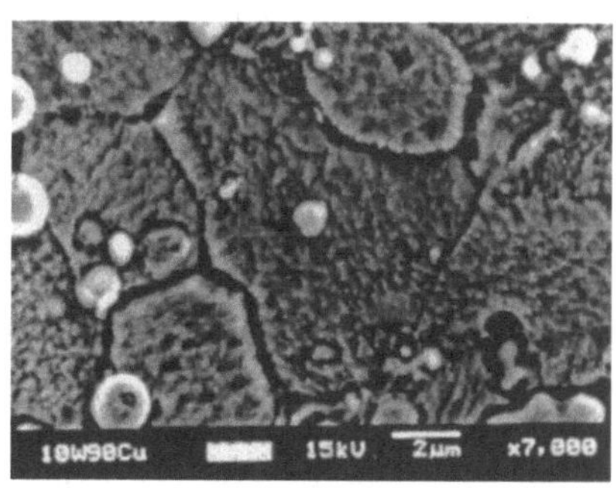

(b)

Fig. 6.34 Microstructure of Cu-W (*P* =1000W, *V* =200mm/min, δ =3.5mm): (a) Cu40wt.%W and (b) Cu10wt.%W.

In general, if W particles can be separated by Cu, the porosity can be dramatically reduced. Very good integrity between Cu and W can be seen, as shown in Fig. 6.35. No debonding or crack can be seen. The wetting behavior of W by molten Cu is greatly affected by the atmosphere [37]. A contact angle of zero degree can be obtained in pure, dry hydrogen atmosphere. Wetting behavior becomes poorer with increasing oxygen content of the Cu powder[38].

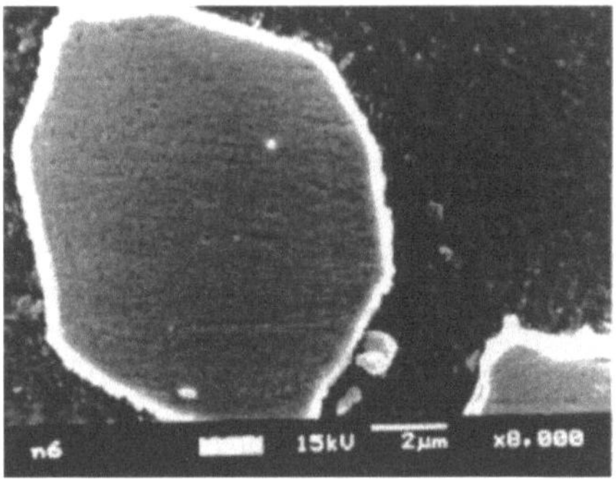

Fig. 6.35 Interface between W particle and Cu matrix of Cu40wt.%W (P = 1000W, V = 200mm/min, δ = 3.5mm).

There is a danger that contamination may occur during milling of W with Cu if mechanical milling is used. Since the W particles are extremely harder compared with most milling jar, Fe based alloy from the jar can contaminate W particles. This effect is illustrated in Fig. 6.36. Some fine dendritic intermetallic structures of W-Fe-Cr-Co can be observed in the specimen with 10W-90Cu. However, the presence of contamination may help to melt Cu to some extent by reduction of reflectivity and formation of Cu alloy.

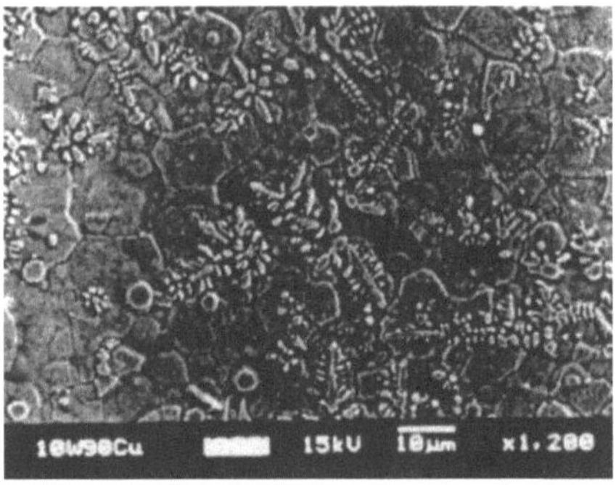

Fig. 6.36 Contamination of Fe and other elements from milling tools.

6.2.3.3 Influence of Ni on Cu-W system

The intrinsic disadvantages of using Cu-W are no solubility between the two constituents, high reflectivity of Cu to laser energy, thermal conductivity of Cu and melting point of W. These effects cause difficulty in melting of Cu by laser and formation of porosity. To improve absorption of the laser energy, the Cu-W material system can be modified by introducing certain elements that can easily absorb the laser energy as well as form diffusional bonding with Cu and W. One of the elements that can be considered is Ni.

Both Ni and Cu have an fcc crystal structure and relatively similar atomic radii. When these two elements are melted during the laser melting process, a Ni-Cu solid solution is formed. The lattice parameters of Ni and Cu are 3.523 Å and 3.6142 Å respectively. The atomic radii for Cu and Ni are 0.128 and 0.125 nm respectively [39]. The lattice parameter of the solid solution decreases as the amount of Ni is increased. The W and Ni phase diagram shows a maximum solubility of 0.3at.% Ni in W at 1495°C while W has about 12.5at.% (31wt.%) solubility in Ni at room temperature [40]. Therefore, Ni is a good element to be used in the formation of the electrode.

After laser scanning, the W diffraction peak shifts slightly, indicating diffusion of Ni into W. However, as the percentage of Ni decreases from 9.5 wt.% to 4.75 wt.%, the shift in the W peak is not so significant. Since Ni is soluble in both W and Cu, it is a good agent to enhance the bonding, and hence densification, through a small increase in the solubility of W in the liquid Cu phase [41]. More detailed enhancement of boning model is proposed in Fig. 6.37, where Ni is located between W and Cu. During laser scanning, Ni first melts and wets with both Cu and W. Melton Ni diffuses into Cu, causing melting of Cu, and into W, resulting in formation of Ni-Cu solid solution as well as Ni-W intermetallics. At the same time Cu and W also solute into Ni, forming compounds.

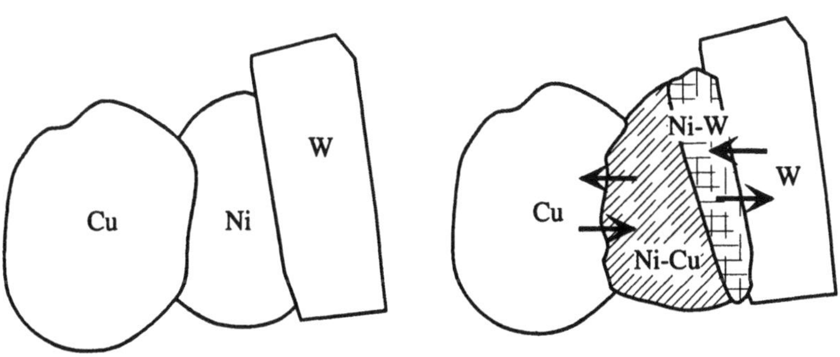

Fig. 6. 37 Mechanisms of enhancement of W and Cu bonding.

The cross-sectional microstructure of Cu50wt.%W9.5wt.%Ni and Cu45wt.%W4.75wt.%Ni specimens are shown in Fig. 6.38 (a) and (b), respectively. Microstructures of both systems show very dense structures, with W particles uniformly distributed throughout the Cu-Ni matrix.

Fig. 6.39 shows EDX of Cu50wt.%W9.5wt.%Ni specimen. The dendritic region and inter-dendritic region consist of 18.2Ni-68.65Cu and 15.58Ni-

71.31Cu, respectively, with some small percentages of W, C and Fe (from the Ni self-fluxing powder).

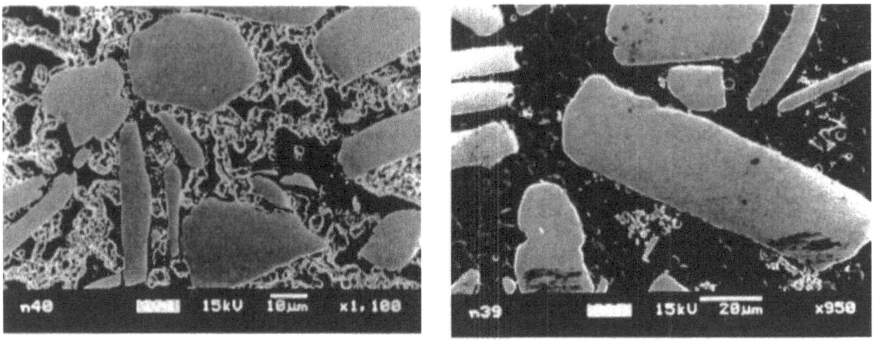

(a) (b)

Fig. 6. 38 Microstructure of Cu-W-Ni: (a) Cu50wt.%W9.5wt.%Ni and (b) Cu45wt.%W4.75wt.%Ni

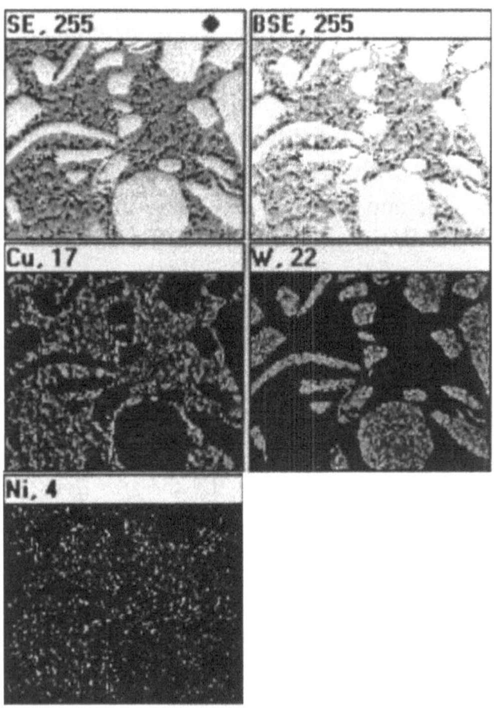

Fig. 6. 39 EDX of Cu50wt.%W9.5wt.%Ni.

Another reason for using Ni as an alloying element, other than increasing the wetting between W and Cu, is because a Ni-Cu powder system is able to form a relatively dense single-line track when the percentage of Ni is varied from 10wt.%-60 wt.%. This is mainly attributed to the high absorptivity of laser energy by Ni and subsequent conduction of heat energy to adjacent Cu particles. Copper powder can be directly melted by the laser beam, as laser light impinging on a powder surface can be reflected from particle to particle, delivering a fraction of the beam power upon each reflection [36]. Once the Cu is in molten state, there is a high tendency for it to "ball", which is caused by surface tension effects [23]. This may lead to uneven surfaces and consequently voids in multi-layered parts. However, with addition of Ni, both Cu and Ni melt, resulting in decrease in the surface tension and formation of a smoother surface.

Fig. 6.40 (a) and (b) show the difference in microstructures of the specimens containing 50wt.%W, with and without thc addition of Ni. The specimen without Ni shows very porous microstructure, whereas with the addition of 9.5wt.%Ni, the pores are almost completely eliminated. The interfacial integrity between the W particles and Cu-Ni matrix is also observed to be relatively good, as shown in Fig. 6.41.

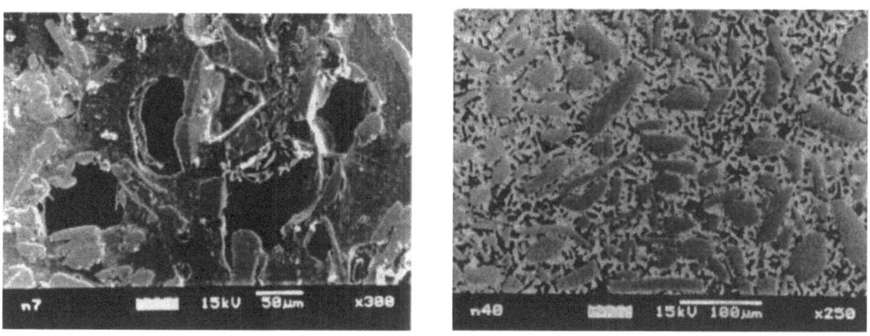

(a) (b)

Fig. 6.40 Cross-sectional microstructure: (a) Cu50wt.%W and (b) Cu50wt.%W9.5wt.%Ni (P = 1000W, V = 100mm/min, δ = 3.5mm)

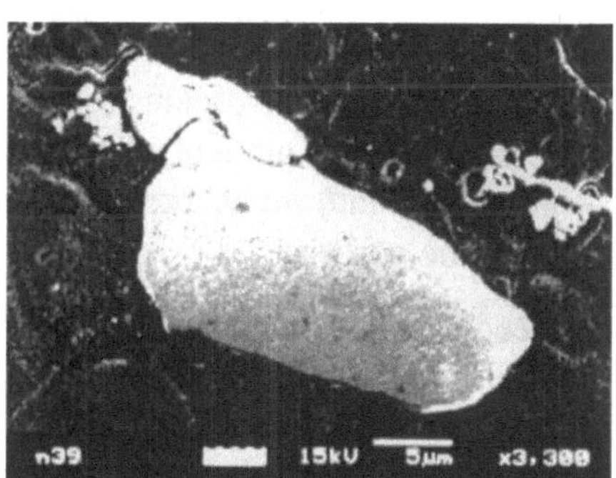

Fig. 6.41 Microstructure of Cu50wt.%W9.5wt.%Ni (P = 1000W, V = 100mm/min, δ = 3.5mm)

It should be noted that a little addition of Ni can also cause an increase in the wetting angle between W and Cu, but the wetting angle decreases slowly with further increase in the amount of Ni [57]. Ihn *et al.* [57] found the change in wetting angle with the amount of additive (Fig. 6.42). As the percentage of Ni is increased beyond 0.2wt.%, the wetting angle progressively decreases. Wetting can also be promoted by the addition of small amount of Co or Fe. Improved wetting is caused by the diffusion of Co and Fe in liquid Cu to form intermetallic compounds, Co_7W_6 and Fe_2W, at the surface of the W particles [57].

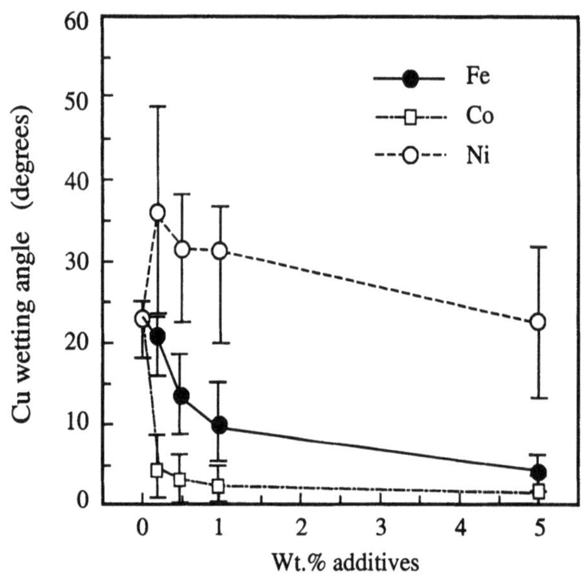

Fig. 6.42 Change in wetting angle of Cu on W with the amount of additive [57].

Fig. 6.43 shows the cross-sectional microstructure of W22.5wt.%-Cu2.38wt.%Ni. As the percentage of W is increased beyond 60 wt.%, the porosity becomes very high. This is mainly attributed to the formation of W frame. As the laser scans the powder bed, only some slight melting of W particles is induced. Therefore, a three-dimensional skeleton of tungsten particles is held together. Because the amount of Cu and Ni is too low, the liquid phase is not enough to flow into the W skeleton. Vaporization of Cu-Ni may also have contributed to the porosity.

The porosity can be reduced if thin powder layer is deposited. If a very thin layer of powder can be laid, the porosity in the initial layer, which has been laser sintered, will be filled up with fine metal powder from the top layer. This will therefore drastically reduce the porosity present. A thinner powder layer also allows for the gaseous impurities to escape more easily from the powder bed.

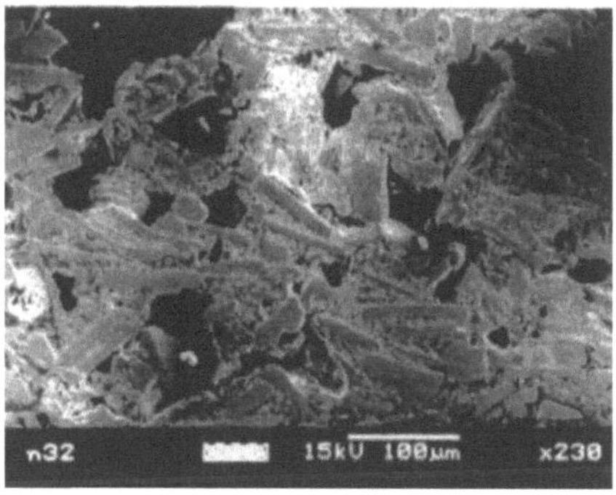

Fig. 6.43 Microstructure of W22.5wt.%Cu2.38wt.%Ni showing the high degree of interconnected porosity (P = 1000W, V = 75mm/min, δ = 3.5mm).

Another possible way to decrease the porosity is to increase the wetting between the W and Cu phase. This can be achieved by coating the W particles with a layer of metallic Ni. In the work of Raghunathan *et. al.* [58] a pre-calculated quantity of hydrous nickel nitrate was mixed with W powder in water and heated to 368 K to evaporate the water, thus leaving a nickel nitrate coating on the W particles. The coated powder was subsequently reduced in a H_2 atmosphere at 873 K, resulting in a coating of metallic Ni.

6.3 Composites

Composites are defined by metal matrix and reinforcements that include carbides, borides, oxides and nitrides, such as SiC, WC, TiC, TiB_2, ZrB_2, ZrO_2, Al_2O_3. Composites have high Young's modulus, yield strength and wear resistance compared to their metal matrix materials.

Metal matrix composites can be produced by ex-situ processes as well as in-situ processes.

6.3.1 Composite system using ex-situ processing

6.3.1.1 Principle of ex-situ process

In ex-situ processing, the reinforcement is ex-situly added into the melt matrix. The metal powder should be first mixed with ceramic particulates that are generally smaller than metal powder particles. Then the mixture of metal and ceramic powders can be laser processed. Due to the difference in the densities between the metal powder and ceramic particulates, segregation is often encountered during the laying of the powder band. To overcome this problem, the powder mixture can be mechanically milled for a few hours. After milling, ceramic particulates can be firmly bonded with metal powder, as schematically shown in Fig. 6.44.

Fig. 6.44 Schemetic view of composite powder particles after mechanical milling of metal powder and ceramic particulates.

Although ceramic particulates can be well mixed with metal powders prior to SLM, settlement of the ceramic particulates due to the density difference may still occur during SLM, especially in the case of large density difference, and high laser power and spot size. The cross-sectional microstructure in Fig. 6.45 shows some degree of segregation of WC particulates in a solid solution of Cu-Ni. The segregation of heavier ceramic particulates can be improved by lowering the laser energy density and laser spot size. It is also important to avoid repeat melting of the solidified area.

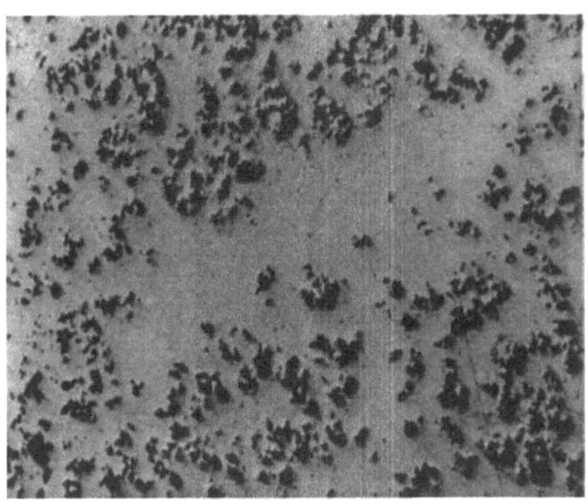

Fig. 6.45. Microstructure of laser processed Cu-Ni/WC composite showing some degree of segregation of WC due to the difference in the densities.

Fig. 6.46 shows clearly that WC segregates at the bottom of the top scanned track (indicated by an arrow). Due to segregation of ceramic particulates, some porosity can be found within this area. This is due to insufficient amount of melt flowing into the ceramic skeleton during the laser scanning process. Reducing the scan speed may improve the presence of the voids by increasing the duration for capillary infiltration while the degree of segregation will increase.

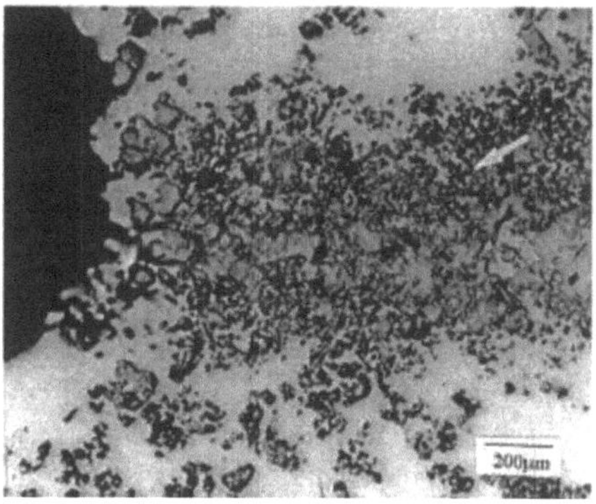

Fig. 6.46 Segregation of WC at the bottom of a single-scanned track.

6.3.2 Composite system using in-situ processing

6.3.2.1 Principle of in-situ reaction

Most composites are fabricated using ex-situ process. The main advantage of using ex-situ process is that it is an easy process as conditions for the synthesis are not so critical. However, since ceramic particulates are incorporated ex-situly, oxidation, and hence formation of thin oxide layer on the surface of the particulates, is unavoidable. Oxide layers on the surfaces of the particulates may affect bonding with metric materials and even react with the metric materials to form weak compounds. In-situ formation of metal-matrix composites is a process where reinforcement phases are formed in-situ during the process.

In general, there are two types of in-situ reactions: gas-solid in-situ reaction and solid-solid reaction.

Fig. 6.47 schematically shows a gas-solid in-situ reaction, where the powder or powder mixture is sealed in a chamber that is filled with reaction gas. Under laser energy, the gas molecule is decomposed. The gas atoms then react with the molten powder to form new phases.

In solid-solid in-situ reaction, two different elemental powders can react with each other under laser heating to form new phases. Several kinds of carbides and borides can be produced by the in-situ reaction from elemental powders:

$$3Fe + C \rightarrow Fe_3C \quad (6.3)$$
$$Ti + C \rightarrow TiC \quad (6.4)$$
$$W + C \rightarrow WC \quad (6.5)$$
$$Ti + 2B \rightarrow TiB_2 \quad (6.6)$$

Besides the reaction between different elemental powders, a compound can also react with another compound or with elemental powder.

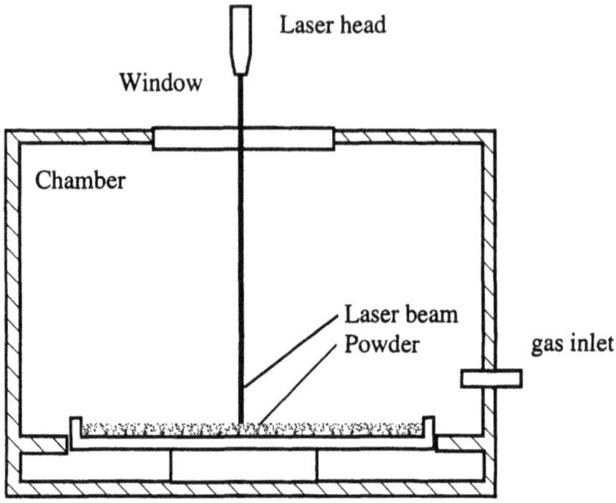

Fig. 6. 47 Principle of gas-solid reaction.

6.3.2.2 Formation of TiC via element reaction

In-situ TiC reinforcement particulates can be formed by direct reaction between elemental Ti and graphite powders. The Ti and graphite powders should be mixed or ball milled with a base powder before laser melting. Since Ti is very reactive at high temperature with many materials, this reaction is easily obtained. One of examples is Cu-TiC composite as given below.

XRD spectra of Cu-Ti-C system are shown in Fig. 6.48. Nominal composition of the system is Cu40wt.%TiC. Before SLM, the Ti and Cu diffraction peaks can be clearly distinguished while some of C diffraction peaks overlap the Cu peaks (Fig. 6.48 (a)). At high melting temperature, Cu and Ti can be mutually soluted. Solubility of Cu in Ti reaches maximum at 1005°C where 12.5 wt.% of Cu can be soluted in Ti while solubility of Ti in Cu is only 7.5at.% at 885°C. Based on the binary Cu-C phase diagram, the solubility of C in Cu is only 0.04 at.% at 1100°C. At room temperature, there is almost no solubility of C in Cu. Therefore only two possible compounds of Cu-Ti and TiC can be formed during melting. Fig. 6.48 (b) shows the XRD spectrum of SLM-processed material. The presence of TiC (111), (200), (220) and (311) diffraction peaks can be clearly seen, indicating formation of in-situ TiC ceramics.

Although this in-situ process is protected by Ar gas, oxidation of C is unavoidable, either from residual oxygen in the glove box or from reduction of oxide layer of Ti. Therefore, this process can reduce the oxidation from oxygen. At the same time, during the in-situ reaction, Cu powder is melted either by conduction heat from Ti or by heat generated from in-situ reaction. No elemental Ti peak can be observed after SLM. Since C may not be able to fully react with Ti, small amount of Ti may react with Cu to form Cu-Ti intermetallic compound. Traces of $CuTi_2$ (006), (204) and (213) diffraction peaks reveal the presence of $CuTi_2$ compound.

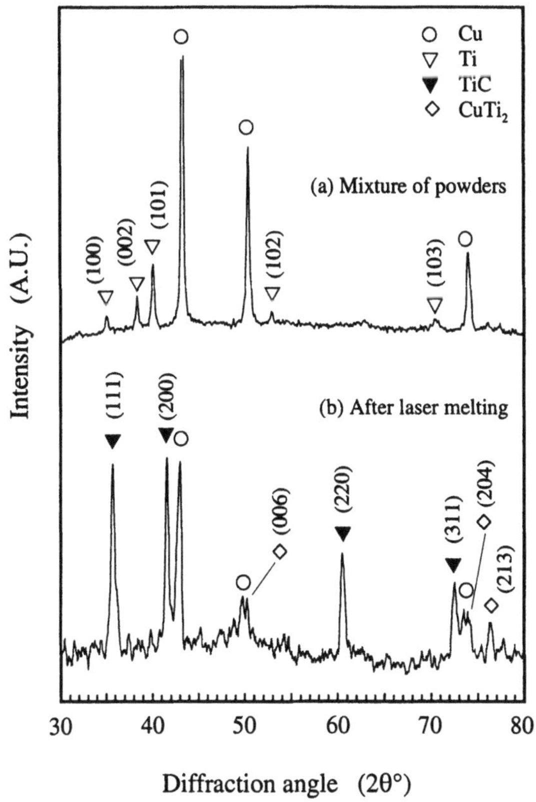

Fig. 6.48 Formation of in-situ TiC in Cu matrix [42].

The possible reaction and accompanying free-energy change for the formation of TiC is [43]

$$\Delta G^\circ = -186606 + 13.22T \quad [\text{Jmol}^{-1}] \tag{6.7}$$

Based on Lu *et al.* [44], The changes in enthalpy and Gibbs free energy can be divided into two temperature ranges:

When $T < 1939$ K,

$$\Delta H = -184571.8 + 5.042T - 2.425*10^{-3}T^2 - 1.958*10^6/T \tag{6.8}$$

$$\Delta G = -184571.8 + 41.382T - 5.042T\ln T + 2.425*10^{-3}T^2 - 9.79*10^5/T \tag{6.9}$$

and when $T \geq 1,939$ K,

$$\Delta H = -160311.5 + 24.79T - 2.732*10^{-3}T^2 - 1.862*10^6/T \tag{6.10}$$

$$\Delta G = -160311.5 - 186.97T - 24.79T\ln T + 2.732*10^{-3}T^2 - 9.31*10^5/T \tag{6.11}$$

It can be seen that the Gibbs energy is always negative, indicating the possibility of formation of TiC.

For Al-Ti-C system, in-situ TiC formation only takes place at about 1191K [45]. In laser melting, the temperature is much higher than this temperature, which is favorable for the formation of TiC. From Lu's calculation [44], the change in temperature due to the reaction heat is 3210K. Because CO_2 laser is unable to directly melt Cu, the melting of Cu is mainly due to conducting heat from Ti and the heat generated by formation of TiC.

Although TiC can be formed, the surface quality of solidified Cu is relatively poor. A large amount of porosity can be easily found in the microstructure. One main cause is vapor from evaporation of graphite during SLM, forming pores subsequently. In addition, due to poor wettability between Cu and in-situ TiC, some porosity is present after SLM.

6.3.2.3 Influence of addition of Ni

Ni powder can be added to the Cu-TiC system to improve the microstructure . The introduction of Ni can increase wettability between Cu and TiC particulates, and also improve the melting behavior of Cu by formation of Cu-Ni solution.

Fig. 6.49 shows the XRD spectrum of Cu-Ni-Ti-C mixture. The nominal compositions of Ni and TiC are 20%. The presence of (111), (200), (220) and (311) TiC diffraction peaks indicates the formation of TiC. Beside Cu-Ni solid solution and TiC phase, some weak (006), (110) and (105) Cu_3Ti_2 diffraction peaks are also found. No $CuTi_2$ phase is observed. This is due to the low percentage of Ti present. No binary Ni-Ti phase has been detected.

Microstructure of an SLM-processed part shown in Fig. 6.50 reveals the distribution of TiC in Cu10wt.%Ni/10wt.%TiC system. TiC particulates of less than 1 μm can be seen (Fig. 6.51). All TiC particulates are either in rectangular or irregular shapes. The TiC particulates are well bounded with Cu matrix. Although some areas show more TiC particulates than others, the distribution of TiC particulates is generally homogeneous. The size of in-situ TiC looks slightly bigger in the Cu20wt.%Ni/10wt.%TiC system than that in the Cu10wt.%Ni/10wt.%TiC system. A large α_2 structure and some defects, in terms of voids, can be observed in the structure. In general, the voids are surrounded by TiC particulates. This suggests the formation of voids due to vaporization of the graphite during SLM. Due to a fast cooling rate of the melt of Cu, the graphite vapor was unable to escape before the solidification of Cu, leading to the formation of voids.

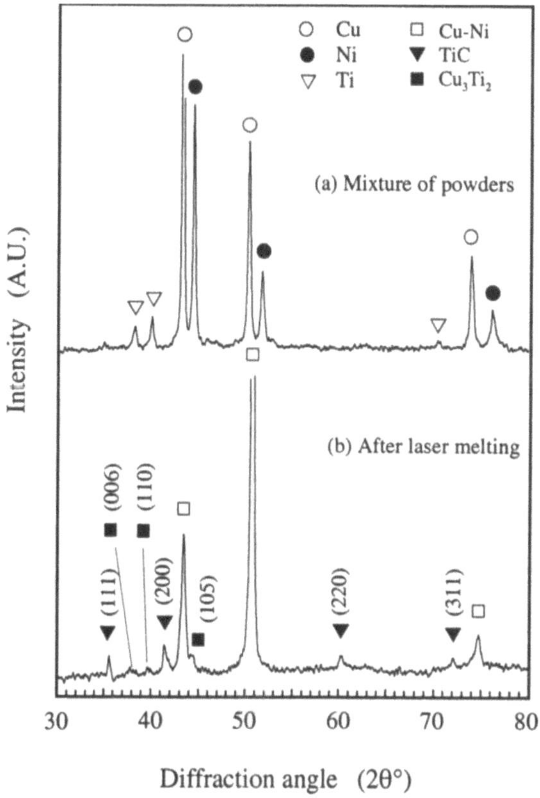

Fig. 6.49 Formation of in-situ TiC in Cu20%Ni matrix [42].

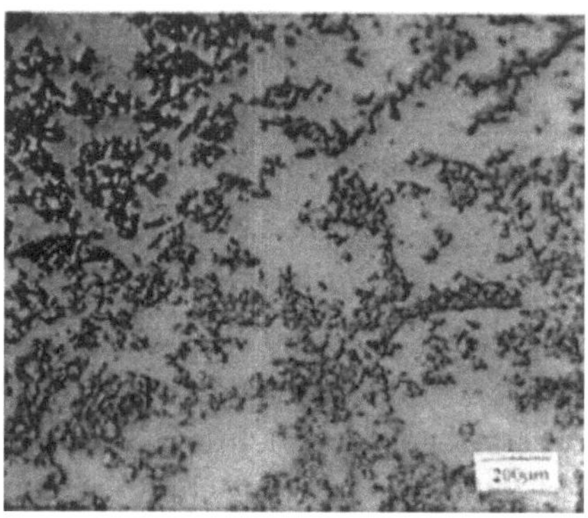

Fig. 6.50 Microstructure of in-situ TiC in Cu10wt.%Ni alloy.

With addition of Ni in this system, porosity in the SLM-processed part can be minimized although some pores can still exist, as shown in Fig. 6.51.

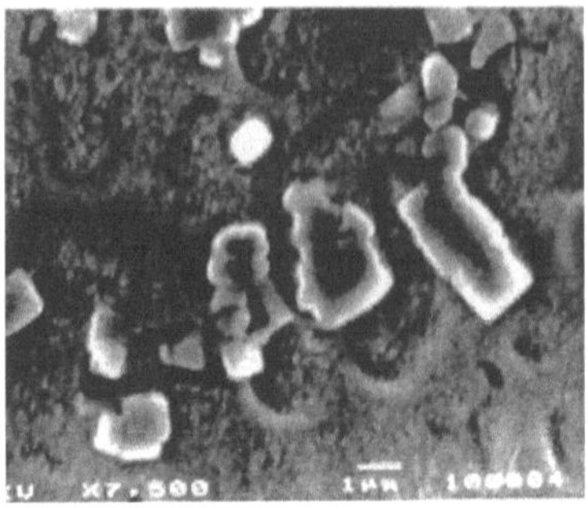

Fig. 6.51 Microstructure of Cu8wt.%Ti2wt.%C15.3wt.%Ni

6.3.2.4 Formation of TiC and TiB_2

There are some difficulties in relying on direct reaction between elements. Graphite is burnt away at high laser power before reacting fully with Ti. This phenomenon leads to formation of porosity and inadequate of graphite with Ti. One way to solve this problem is to employ displacement reaction, such as reaction between Ti and B_4C. However, some extra product, which is TiB_2, will be spontaneously synthesized.

$$3Ti + B_4C \rightarrow 2TiB_2 + TiC \tag{6.12}$$

The above reaction occurs readily to form the final products as the reaction is very exothermic, with $\Delta H_{298} = -686.7$ kJ/mol [46].

Fig. 6.52 shows XRD diffraction spectra of Cu14.4wt.%Ti5.6wt.%B_4C before and after laser fabrication. Since B_4C peak is too week, compared with Ti and Cu diffraction peaks, it does not appear in Fig. 6.52 (a). It is obvious that titanium diboride and titanium carbide are formed after laser melting. The lattice parameter of the in-situ formed titanium carbide is found to be 4.2854 Å, whereas stoichiometric TiC with a fcc crystal structure has a lattice parameter of 4.311 Å. Therefore non-stoichiometric titanium carbide, TiC_{1-x} has been formed when the laser beam scans the powder bed. Capaldi and Wood [47] gave the lattice parameter for $TiC_{0.55}$ to be 4.283 Å; thus, the value of x for the sub-stoichiometric titanium carbide formed is greater than 0.45.

The structure of stoichiometric compound TiC possesses an fcc lattice, in which the carbon atoms are inserted inside the octahedral interstices. Studies have shown that up to 72 percent of interstitial carbon sites can remain vacant. The probable cause of lattice contraction could be a high vacancy content inside the TiC lattice due to carbon vacancies generated from the fast solidification, and stabilized by the subsequent cooling [48]. Thus, carbon atoms have insufficient time to diffuse into the TiC_{1-x} lattice to form stoichiometric TiC due to the short laser melting time.

$$xC + TiC_{1-x} \rightarrow TiC \tag{6.13}$$

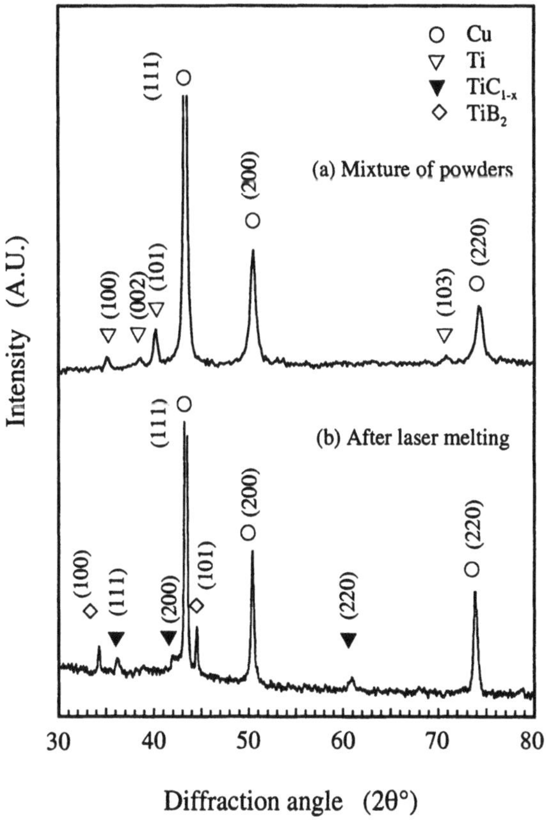

Fig. 6.52 XRD spectra of Cu14.4wt.%Ti5.6wt.%B_4C: (a) before laser melting and (b) after laser melting.

The formation of sub-stoichiometric titanium carbide is not uncommon, as titanium carbide has an extremely wide range of homogeneity [49]. Brodkin *et al.* [50] reported the formation of $TiC_{0.5}$ as an intermediate phase when a starting powder composition of Ti/B_4C in the ratio 4:1 was hot pressed. The stoichiometry of TiC_{1-x} changed from 0.5 to 0.65 as the sample was soaked at 1600°C for 4 hours, thereby increasing the number of carbon atoms diffusing into the TiC crystal lattice. In another investigation by Fouilland-Paille *et al.* [48], non-stoichiometric TiC_x was formed when a laser beam irradiated a coating of Ti and C.

The presence of oxygen may be another reason for the reduction in lattice parameter of TiC. Yamada [51] reported that it is possible for oxygen, contained in the powder and the reaction cell, to substitute the carbon atoms in the TiC lattice and consequently reduce the lattice parameter.

The lattice parameter of the sub-stoichiometric TiC for Cu7.22wt.%Ti2.78 wt.%B_4C and Cu14.44 wt.%Ti5.56 wt.%B_4C, both scanned with the same laser parameter (P=200W, V=300mm/min, δ=3.5mm), are 4.2169 and 4.2535 Å, respectively. With reduction in the percentage of Cu, the lattice parameter can be increased. This is attributed to shorter mean diffusional path for the carbon atoms with reduced percentage of Cu, as the Ti and B_4C particles are less separated and more carbon atoms are able to diffuse into the Ti lattice.

XRD shows a slight shift of the Cu diffraction peaks. The lattice parameter of Cu increases to 3.6174 Å (a_{Cu} = 3.6146 Å). EDX spectrum reveals small amounts of C, Fe and Ti in Cu base. The presence of C is due to dissociation of B_4C, whereas Fe may come from the base material during the high-temperature laser processing, or introduced during the ball milling process. Small amount of Ti is soluble in Cu to form a solid solution. The maximum solubility of Ti in Cu is approximately 5 wt.% at 885°C. Therefore, it is most likely that larger Ti atoms of an atomic radius of 0.1446 nm has substituted Cu atoms (0.1278 nm) in the crystal lattice, causing the crystal to expand.

Some small amount of B_4C may remain un-reacted as a certain percentage of Ti is in solution with Cu. This amount may not give detectable peaks in the x-ray diffraction spectra, as x-ray is insensitive to small amount of B_4C. No compounds of Cu and Ti were detected due to the non-equilibrium and very short laser sintering time.

Brodkin *et al.* [50] reported the formation of Ti_3B_4 when the starting powders of 4:1 Ti/B_4C were compacted, sintered and soaked at 1600°C for 4 hours. At this temperature and condition, TiB reacted with TiB_2 to form large platelets like Ti_3B_4. Udwadis et al. [52] reported that when a powder mixture with Ti to B_4C ratio 3:1 was reacted by the combustion synthesis method, only TiB_2 and TiC phases formed from the B_4C powders with average particle diameter of less than 8 μm, and Ti with average particle diameter of less than 55 μm. For coarser Ti powders, other phases, such as Ti_3B_4 and TiB, were present in the product, especially when coarser B_4C powders were used. The starting average particle size for B_4C and Ti used in the experiment were 2 μm and <150 μm, respectively. For all the powder batches for this particular powder system, neither TiB nor Ti_3B_4 was detected in the final product, as was reported by the other researchers. The possible reason could be due to the different processing technique employed.

In laser melting, when the laser scans the powder bed, the laser energy is absorbed by the powder particles, followed by transformation to liquid state. The melting point for Ti and Cu are 1600°C and 1083°C, respectively. Some Ti solutes with Cu and some reacts with B_4C at the B_4C and Ti melt interface. The products of the reaction, TiB_2 and TiC_x, do not react with Cu since they are not soluble in Cu. The formation of a liquid phase during laser sintering promotes the in-situ reaction as the interaction between the particles is increased.

Possible reaction routes for the direct reaction between Ti and B_4C have reported by Barsoum *et al.* [54], whose samples, consisting of Ti and B_4C in a 3:1 ratio, were hot pressed. To facilitate discussion, TiC rather than TiC_{1-x} is used, as no thermodynamic data is available for sub-stoichiometric titanium carbide.

The reaction mechanism of Cu20wt.%(Ti33wt.%B_4C) was also investigated using a high temperature DSC [53]. Two thermal cycles from room temperature to 1200°C were performed. Fig. 6.53 shows the DSC results. Three exothermic and one endothermic peaks have been detected at 580, 820, 990, and 1090°C, respectively. Obviously the endothermic peak at 1090°C was due to the heat release of Cu melting. To understand these thermal events, the specimens heat treated at 600, 750, 900, 1050 and 1200°C were investigated using XRD. Six XRD diffraction spectra are given in Fig. 6.54.

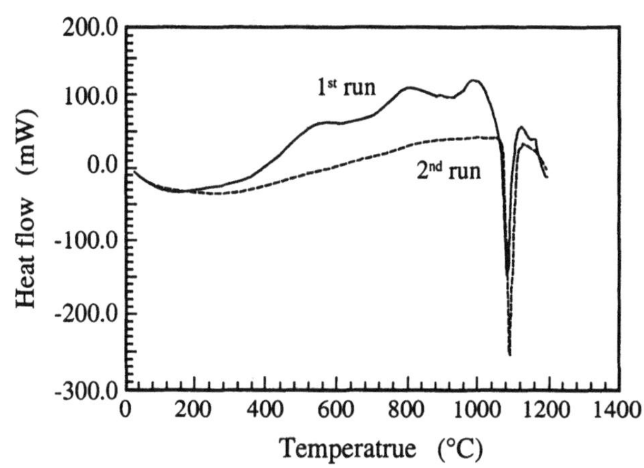

Fig. 6.53 DSC measurement of Cu14.4wt.%Ti5.6wt.%B_4C.

A new phase of Cu_4Ti can be seen from Fig. 6.54 when the powder is heat-treated at 600°C. Discovery of Cu_4Ti indicates that the peak at 580.19°C is due to formation of Cu_4Ti by reaction between Ti and Cu. With increase in temperature to 750°C, the intensity of the Cu_3Ti intermetallic phase decreases, indicating that this phase may not be stable at elevated temperature. There is no clear indication of the formation of new phases when the specimen is heat-treated at 750°C.

For the specimen treated at 900°C, Ti_2B_5–like phase can be detected, indicating that the second exothermic peak at 817°C is due to formation of Ti_2B_5. Some other traces of TiB and TiC_{1-x} can be seen, but they are not conclusive. During this stage of reaction, B_4C is dissociated, releasing B and C atoms to react with Ti to form TiB, Ti_2B_5 and TiC, according to Eq. (6. 14).

This is in agreement with the work by Barsoum *et al.* [54] in which Ti_2B_5, TiB and TiC_x phases were detected when a sample with a 3:1 Ti/B_4C ratio was heated in the DTA.

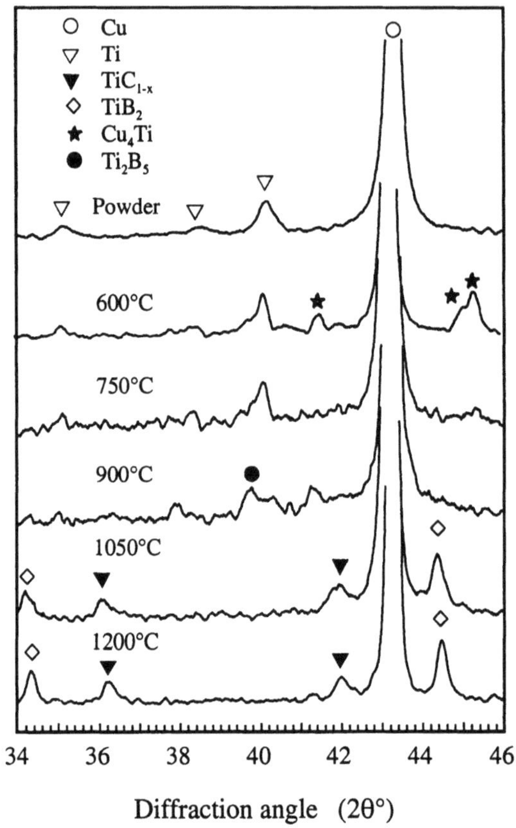

Fig. 6.54. XRD spectra of Cu14.4wt.%Ti5.6wt.%B_4C heated to different temperatures.

TiB and Ti_2B_5 phases completely disappear after the sample is heated to a temperature of 1050°C, indicating decomposition of Ti_2B_5. According to Barsoum *et al.* [54], TiB subsequently reacts with Ti_2B_5 to form TiB_2 as in Eq. (6.15), and this corresponds to the third exothermic peak at 996°C. The net reaction results in the formation of TiB_2 and TiC (Eq. (6.16)).

$$3Ti + B_4C \rightarrow (2/3)TiB + (2/3)Ti_2B_5 + TiC \quad (6.14)$$

$$(2/3)TiB + (2/3)Ti_2B_5 \rightarrow 2TiB_2 \quad (6.15)$$

$$3Ti + B_4C \rightarrow 2TiB_2 + TiC \quad (6.16)$$

The lattice parameter of the in-situ formed titanium carbide is 4.2901 Å; therefore, non-stoichiometric TiC_{1-y} is formed. The relatively high heating rate of 60°C/min affects the formation of titanium carbide. The lattice parameter of the sub-stoichiometric titanium carbide formed by laser sintering is smaller compared to that heat-treated in high-temperature DSC.

It should be noted that this might not be the actual reaction path for the laser sintering process of a powder system consisting of Ti-B_4C-Cu. The heating rate and the maximum temperature attainable with the laser are very much higher compared to the DSC test. Furthermore, no literature has been found to substantiate the reaction mechanism of this particular powder composite.

Another possible reaction route for the Ti-B_4C reaction has been reported by Mogilevsky *et al.* [55] and Brodkin *et al.* [50], who studied the interaction between B_4C and Ti coatings and reactive hot pressing of Ti-B_4C composites, respectively. For the calculation of the change in Gibbs free energy, the actual reaction between B_4C and Ti in molten Cu can be expressed by:

$$x\text{Ti} + y[\text{Ti}]_{\text{Cu}} + \text{B}_4\text{C} \rightarrow 2\text{TiB}_2 + \text{TiC}$$
$$\Delta G = \Delta G^\circ + 2RT\ln(a_{\text{TiB2}}) + RT\ln(a_{\text{TiC}}) - xRT\ln(a_{\text{Ti}}) - yRT\ln(a_{[\text{Ti}]\text{Cu}})$$
$$\Delta G^\circ = 2\Delta G_{\text{TiB2}} + \Delta G_{\text{TiC}} - x\Delta G_{\text{Ti}} - y\Delta G_{[\text{Ti}]\text{Cu}} \qquad \text{at } 25°\text{C}$$

where y is moles of Ti dissolved in Cu, x, moles of Ti, R, gas constant, T, temperature, and a, activity.

During the initial stages of the reaction, C diffuses rapidly from B_4C particles into Ti, leaving behind areas of high B activity [50]. The diffusion of C in Ti is significantly greater than B. Boron and carbon then react with Ti to form TiB and TiC. Due to the sluggish diffusitivity of B, TiB is expected to form near the original B_4C field. Therefore at the initial stages, TiB and TiC are formed according to Eq. (6.17). A small amount of TiB and TiC was detected from the STA test at 900°C, as shown in Fig. 6.53.

$$5\text{Ti} + \text{B}_4\text{C} \rightarrow 4\text{TiB} + \text{TiC} \qquad (6.17)$$
$$\text{B}_4\text{C} + 2\text{TiC} \rightarrow 2\text{TiB}_2 + \text{C} \qquad \Delta G = -197 \text{ kJ/mol} \qquad (6.18)$$
$$\text{B}_4\text{C} + 4\text{TiB} \rightarrow 4\text{TiB}_2 + \text{C} \qquad \Delta G = -516 \text{ kJ/mol} \qquad (6.19)$$

A plot of the change in Gibbs free energy in Fig. 6.55 shows that TiB forms in favor of TiB_2 at the initial stages of reaction.

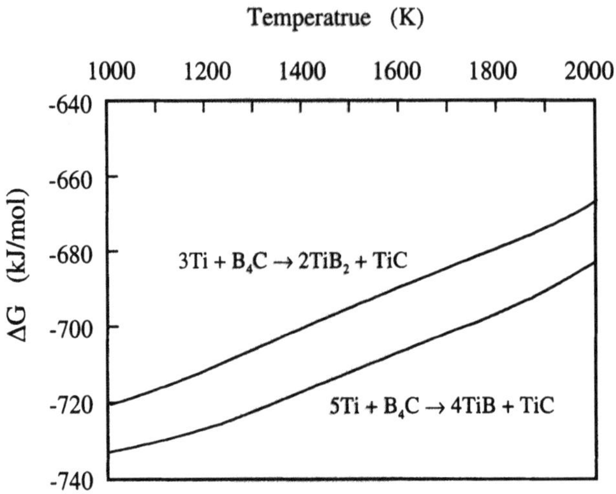

Fig. 6.55 Change in Gibbs free energy.

TiB is not stable when in contact with B_4C and decomposes rapidly to form TiB_2. The change in Gibbs free energy is negative, indicating that the reaction is favorable (see Eq. (6.18) and (6.19)). A schematic of the reaction sequence between Ti and B_4C is depicted in Fig. 6.56. Some residual TiB might be present, although no TiB is observed from XRD analysis due to the low amount present.

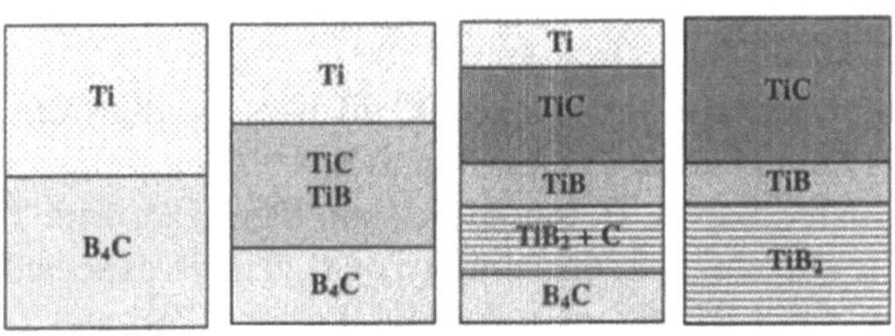

Fig. 6.56 Schematic diagram of the reaction between Ti and B_4C.

6.4 Summary

Selective laser melting (SLM) and laser cladding techniques may enable advances in fabrication of fully dense structure, with corresponding increased mechanical properties. By introducing ceramic reinforcement into the metal matrix, wear resistance can be improved. Due to full melting of the powder, the accuracy of selective laser melting and laser cladding is relatively lower in comparison with selective laser sintering. Process parameters should be accurately controlled in order to minimize distortion. Balling in SLM is a major draw-back. Therefore wetting and protection of oxidation of the powder are crucial.

6.5 References

1. M. Agarwala, D. Bourell, J. Beaman, H. Marcus and J. Barlow, *RP. J.*, Vol. 1 (1995), 26.
2. O.D.D. Soares and M. P. Amor, *Applied Laser Tooling*, Martinus Nijhoff, 1987.
3. W.M. Steen, *Laser Material Processing*, Springer-Verlag, 1991.
4. J. Gerken, H. Haferkamp and H. Schmidt, *Proc. 27th ISATA*, Achen, Germany, Automotive Automation, Croydon, 1994.
5. S.C. Peng, C.J. Chou and J.Y. Jeng, *Proc. 27th ISATA*, Achen, Germany, Automotive Automation, Croydon, 31Oct. – 4 Nov. 1994, 273.
6. V.M. Weerasinghe and W.M. Steen, *Computer simulation model for laser cladding, Transport Phenomena in Mater. Proc.*, New York, ASME, 1983, 15.
7. T.R. Anthony and H.E. Cline, *J. App. Phys.*, Vol. 48 (1977), 3888.
8. C. Chan, J. Mazumder and M.M. Chen, *Lasers in Mater Proc*, ASM, 1983, 150.
9. R. Vilar, *Mater. Sci. Forum*, Vol. 301 (1999), 229.
10. V.M. Weerasinghe and W.M. Steen, *Applied Laser Tooling*, Ed. O.D.D. Soares and M. Perez-Amor, Martinus Nijhoff Pub., Dordrecht, 1987, 183.
11. Z.D. Chen, T.A. Mai and G.C. Lim, *Conf. Proc. of ICALEO'98*, Orlando, Florida, USA, 1989.
12. P.A. Vetter, T. Engle and J. Fontaine, *Laser Mater. Proc.: Industrail and Microelectronics Applications*, Bellingham: Spie – Intern. Soc. Optical Eng., 1994, 452.
13. J. Lin and W.M. Steen, *J. Laser Applications*, Vol. 10 (1998), 55.
14. J.Y. Jeng, S.C. Peng and C.J. Chou, *Intern. J. Adv. Manuf. Tech.*, Vol. 16 (2000), 681.
15. N.K. Tolochko, T. Laoui, Y.V. Khlopkov, S.E. Mozzharov, V.I. Titov and M.B. Ignatiev, *RP J.*, Vol. 6 (2000), 155.
16. F.G. Arcella and F.H. Froes, *JOM*, Vol. 52, (5) (2000), 28.
17. D.H. Abbott and F.G. Arcella, *Adv. Mater. Proc.*, Vol. 5 (1998), 29.
18. G. K. Lewis and E. Schlienger, Mater. Design, Vol. 21 (2000), 417.
19. J. Mazumder, J. Choi, K. Nagarathnam, J. Koch and D. Hetzner, *JOM*, May (1997), 55.
20. J.A. Benda, Solid Free Form Fabrication Symp., The University of Texas at Austin, 1994, 277.
21. Y. Song, *CIRP Ann.* 1997, Manufacturing Tech., Pub. Annl. Inter. Inst. for Prod. Eng. Res., Vol. 46/1 (1997), 129.
22. H.J. Niu and I.T.H. Chang, Scripta Mater., Vol. 41 (1999), 1229.

23 D. L. Bourell, H.L Marcus, J.W. Barlow, J.J. Beaman, Intern. *J. Powder Metall.*, Vol. 28, No. 4 (1992), 371.

24. B.V. Schueren and J.P. Kruth, *Laser Assisted Net Shape Eng. Proc. of the LANE'94*, Ed. M. Geiger and F. Vollertsen, Meisenbach Bamberg, Erlangen, Vol. 11 (1994), 793.
25. L. Rayleigh, Proc. London Math., Vol. 10, (1878), 4.
26. F.A. Nichols and W.W. Mullins, Trans. Met. Soc. AIME, Vol. 233 (1965), 1840.
27. H.J. Niu and I.T.H. Chang, Scripta Mater., Vol. 39 (1998), 67.
28. K.C. Mills and B.J. Keene, Inter. Mater. Rev., Vol. 35 (1990), 185.
29. H.J. Niu and I.T.H. Chang, J. Mater. Sci., Vol. 35 (2000), 31.
30. F. Costache, A. Marian, D.M. Buca and V. Iov, *SIOEL'99, Sixth Symp. on Optoelectronics*, Ed. Teodor Necsoiu, Maria Robu, Dan C, Dumitras, Proc. of SPIE Vol. 4068 (2000), 555.
31. F.H. Froes, *Mater. Tech.*, Vol. 15 (2000), 8.
32. P.K.D.V. Yarlagadda, P. Christodoulou and V.S. Subramanian, *J. Mater. Proc. Tech.*, Vols. 89-90 (1999), 231.
33. G. Semon, *A Practice Guid to Electro-Discharge Machining*, 2nd ed., Ateliers des Charmilles, Geneva, Chapter 9, 1975, 63.

34. B.E. Stucker, W.L. Bradley, S.Norasetthekul and P.T. Eubank "*The Production of Electrical Discharge Machining Electrodes Using SLS : Preliminary Results*", Texas A&M University.
35. A. Upadhyaya, R.M. German, Intern. *J. Powder Metall.*, Vol. 34, (2) (1998), 43.
36. S S. Charshan, *Guide to Laser Materials Processing*, Laser Institue of America, 1993.
37 N.M. Parikh, M. Humenik, *J. Amer. Ceramic Soc.*, Vol. 40 (1957), 315.
38 M. Nicholas and P.M. Pools, *J. Mater. Sci.*, Vol. 2 (1967), 269.
39 W.D. Callister, *Mater. Sci. Eng.*, 2nd Edition, John Wiley & Sons, 1991, 74.
40. *Metal Handbook Binary Alloy Phase Diagrams*, 2nd edition ASM International: Materials Park, OH., Vol. 2 (1991), 1442.
41. M.M. Sun, J.C. Nelson, J.J. Beaman, J.W. Barlow, *Solid Freeform Fabrication Proceedings*, The University of Texas, Austin, Texas, 1991, 46.
42. L. Lu, J.Y.H. Fuh, Z.D. Chen, C.C. Leong and Y.S. Wong, Mater. Res. Bul., Vol. 35 (2000), 1555.
43. R.G. Colters, *Mater. Sci. Eng*, Vol.70 (1985), 1.
44. W.J. Lu, X.N. Zhang, D. Zhang, R.J. Wu, Y.J. Bian an P.W. Fang, *Acta Metall. Sinica*, Vol. 35 (1999), 536.
45. T. Nukami and M. C. Flemings, *Metall. Mater. Trans.* A, Vol. 26A (1995), 1877.
46. F.D.S. Marquis, S.S. Batsanov, J.A. Puszynski, *Proc. and Fabrication of Adv. Mater. V*, The Minerals, Metals & Materials Soc, 1996, 668.
47. M.J. Capaldi, J.V. Wood, *J. Mater. Synth. Proc.*, Vol. 4, (4) (1996), 245.
48. L. Fouilland-Paille, S. Ettaqi, S. Benayoun, J.J. Hantzpergue, *Surface Coatings Tech.*, Vol. 88 (1996), 204.
49. I. Gotman, M.J. Koczak, *Mater. Sci. Eng.*, A187 (1994), 189.
50 D. Brodkin, Surya R. Kalidindi, Michel W. Barsoum, A. Zavaliangos *J. Amer. Ceram. Soc.*, Vol. 79, (7) (1996), 1945.
51 O. Yamada, *J. Amer. Ceram. Soc.*, Vol. 70, (9) (1987), C206.
52 K.K. Udwadia, J.A. Puszynski, *In-situ Reactions for Synthesis of Composites, Ceramics and Intermetallics*, edited by E.V. Barrera, F.D.S. Marquis, S.G. Fishman, N.N. Thadhani, W.E. Frazier, Z.A. Munir, TMS, 1995.
53. C.C. Leong, *M.Eng Thesis*, National University of Singapore, (2000).
54. M.W. Barsoum, B. Houng, *J. Amer. Ceram. Soc.*, Vol. 76, (6) (1993), 1445.
55. P. Mogilevsky, J.Gotman, E.Y. Gutmanas, *Mater. Sci. Eng.*, A171 (1993), 271.
56. D.L Bourell, H.L. Marcus, J.W. Barlow, and J.J. Beaman, *J. Powder Metall.*, Vol. 28, (4) (1992), 371.
57. T.H. Ihn, S.W. Lee, S.K. Joo, *Powder Metall.*, Vol. 37, (4) (1994), 283.
58 S.K. Raghunathan, C.Persad, D.L. Bourell, H.L. Marcus, *Mater. Sci. Eng.*, A131 (1991), 243.

Chapter 7

LASER SINTERING OF CERAMICS

7.1 Fabrication of ceramic parts using SLS

Over the last decade, several RP methods have been developed as techniques to directly or indirectly fabricate ceramic parts. Each process has its own characteristics, advantages and disadvantages. To make a sound decision in the selection of these RP-based processes, one must have a clear understanding of the processes and their respective advantages and disadvantages. These RP processes include laser lithography (LL), fused deposition modeling (FDM), selective laser sintering (SLS) [1-7], laminated object manufacturing (LOM), 3-D printing and Sanders Model Maker ™ etc. Formal efforts have demonstrated the feasibility of these processes for the direct fabrication of functional ceramics [8-11]. In addition, CAM-LEM, shape deposition manufacturing (SDM), multiphase jet solidification (MJS), selective area laser deposition (SALD), laser engineered net shaping (LENS), and extrusion freeform fabrication are being developed for fabrication of ceramic parts [12-14].

LL and LOM are indirect processes to fabricate ceramic parts. Molds are first made by either of these processes and then used for casting of the ceramic parts. The casting of the ceramic parts is similar to the conventional ceramic process. FDM uses a spool to feed filament into a heated extruding head in the same manner as feeding wire into an automatic welder. The product has good dimensional accuracy and surface finish, and fast building speed. However, the material strength is low. Supports may be needed in the fabrication of parts using this process. Using this process, functional ceramic parts can be directly fabricated.

SLS deposits and sinters a layer of heat-fusible powder to form the respective cross-sections of the part. An initial cross-section of the object under fabrication is selectively "drawn" on the layer of powder by laser. The process does not require support generation and thus only straightforward post-processing needed.

This chapter only focuses on the SLS process for the creation of functional ceramics.

7.1.1 Process

In general, ceramics, have extremely high melting temperatures so that they cannot be directly fused together to form parts. In most cases, indirect SLS is employed in the fabrication of ceramics parts from powders [15-17]. The entire process consists of coating polymer onto ceramic powder, laser sintering to form a green part which contains porosity, infiltration of ceramic sol, debinding polymer binder, and finally sintering to high density. Low percentage of binder is desirable since porosity of sintered parts is proportional to the volume of binder. The binder can be coated or mixed. However, the strength of the processed part using a mixture of ceramic and polymer is very low. A strength comparison of bars made from coated and mixed binders is shown in Fig. 7.1 [18]. The big difference in bending strength is caused by inhomogeneous distribution of polymer in the mixed batch, and segregation of polymer after SLS. Fig. 7.1 also indicates that there is an optimal laser energy at which the highest volume melt of polymer can be obtained. Above the optimum, the viscosity of the polymer is too high and the polymer even evaporates at the higher laser energy.

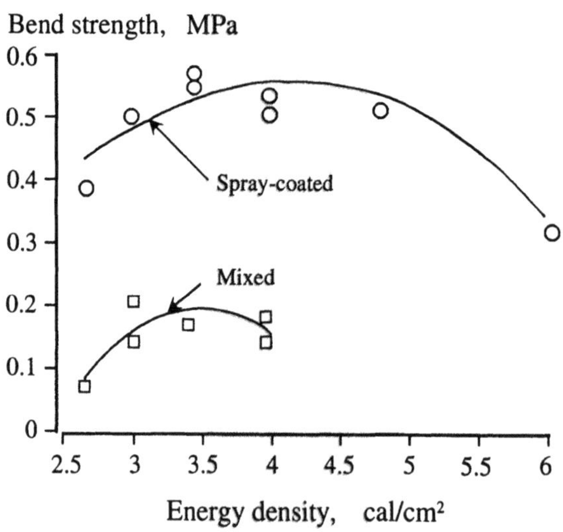

Fig. 7.1 Comparison of strengths of bars made from Al_2O_3 spray-coated or mixed 25vol.% of binder [18].

SLS-processed ceramic parts are very brittle and cannot be used. They should be post-treated to increase their strength. Several post processes can be considered, depending on different ceramic systems. High-temperature sintering is a basic process in which porous ceramic parts can be made denser. However, since there exists very high porosity after SLS, extremely high volume shrinkage will happen during post sintering. Beside that, the porosity percentage is still high even after sintering. To minimize shrinkage and increase sintering density, infiltration of ceramics has been developed by using nanometer ceramic sol. Both density and strength can be improved after sintering the infiltrated parts. Further increase in density and mechanical properties can be achieved by HIP process.

Some novel in-situ formation of ceramics has been developed by using reactive binders and infiltration of metal into the SLS-processed ceramic skeleton. Through chemical reaction or physical solution, the reactive binders become a part of the ceramics, which simplifies the whole processing. Fig. 7.2 summarizes various stages of processing the ceramic parts.

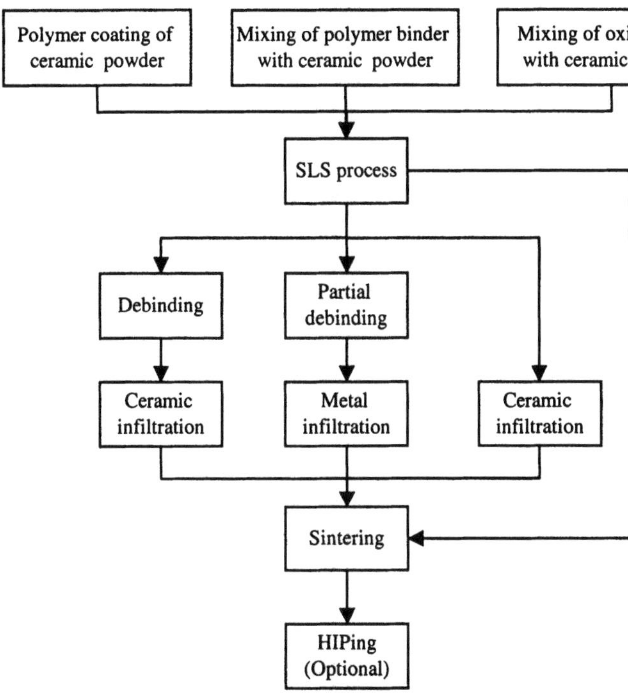

Fig. 7.2 Flow chart for processing ceramic parts.

The amount of binder may strongly affect the mechanical properties of SLS-processed parts. The effect of the binder content on the strength of the composites has been described by Bache [19]. The strength of porous ceramics can be written as:

$$\sigma = k\left(\frac{V_b}{V_p}\right)^{0.75}\sqrt{\left(\frac{\gamma E}{d}\right)} \quad (7.1)$$

where k is a constant, V_b, volume content of binder, V_p, volume content of primary phase, γ, surface fracture energy, E, Young's modulus and d, grain size. The strength of a composite depends on the relative volume of binder and ceramics, grain size of ceramics and Young's modulus as well.

7.2 SLS of ceramic parts

7.2.1 SLS with infiltration

Subramanian *et al.* [18] has used a mixture of Al_2O_3 with average particle size of 15 μm and polymer binders of either poly methyl methacrylate (PMMA) or poly methyl methacrylate-*co*-*n*-butyl methacrylate (copolymer). The flexural strengths in three-point bending of PMMA and copolymer are respectively 21±3 MPa and 35±5 MPa [18]. These two polymers can be thermally decomposed when they are heated above 400°C [20, 21].

The process starts with coating of the polymer to Al_2O_3 powder surface using a alumina-to-polymer volume ratio of 50:50 (for more details, refer to references [22, 23]). CO_2 laser can be used for the SLS process. To minimize warpage and curl, the powder bed should be warmed to 80°C if copolymer is used, and 90°C with PMMA binder. Colloids with nanometer Al_2O_3 must be used for the infiltration process. The green ceramic part after SLS can be infiltrated in Al_2O_3 sol at room temperature by immersing the part into the sol. Since the Al_2O_3 sol only contains about 12% of Al_2O_3, the infiltrated parts should be dried thoroughly in an air oven at 120°C for 24 hours. At this stage, the Al_2O_3 powder particles are still bound by the polymer coating. Therefore the polymer coating should be debound. Debinding should be carried out at a temperature above the decomposition temperature of the polymers. To bind Al_2O_3 particle together, the parts are sintered at a high temperature. Since the parts are very fragile, care must be taken during sintering. The heating rate must be slow to avoid crack due to high internal stresses generated by inhomogeneity of thermal dilatation. It is suggested[18] that several steps of heating should be considered. The parts are firstly sintered at 1050°C at a heating rate of 3.3°C/min for six hours. At this stage of sintering, the nano-sized Al_2O_3 powder from the infiltration sol will be sintered. The sintered parts have certain strength contributed from the infiltrated fine Al_2O_3. There still exists a large amount of porosity and the SLS-processed particles are still separated from one another. Therefore further sintering is necessary. The final sintering temperature is 1,600°C, as for conventional sintering of Al_2O_3, and holding duration is about six hours. The density measurement of sintered Al_2O_3 is given in Fig. 7.3. In general, the laser energy density during SLS does not affect the final sintered density while it does strongly affect the density as well as the bending strength of the SLS-processed parts. It is found that the volume percentage of Al_2O_3 in the colloids contributes to the sintered density. At low volume concentration of Al_2O_3, more pores remain after sintering.

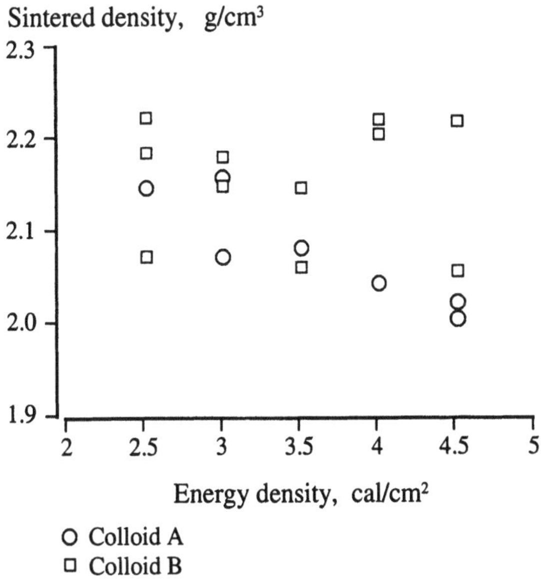

Fig. 7.3. Sintered density of Al_2O_3 parts using different colloids: Colloid A contains 12vol.% of Al_2O_3 with particle size of 60 nm and colloid B contains 16vol.% of Al_2O_3 with particle size of 150 nm [18].

Beside polymer binders, other inorganic binders can also be used, such as monoclinic B_2O_3 [24]. An advantage of using B_2O_3 is its low melting temperature of about 450°C. Boron oxide in borosilicate glasses can reduce the expansion coefficient and improve workability by decreasing the viscosity [25]. The first step of the process is mixing and baking B_2O_3 with Al_2O_3 powder. The resulting HBO_2 that is formed from the B_2O_3 by reacting with moisture and baking has even lower melting point of 170°C, which can be used as an inorganic binder.

The processed HBO_2 is mixed with Al_2O_3 and zinc borosilicate for the SLS process. Lee [26] found that SLS-processed parts made from high content of glass show higher density (Fig. 7.4 (a)) and bend strength (Fig. 7.4 (b)) for all sintering temperature, when compared with those with low glass content. Both density and bending strength increase with sintering temperature until about 900°C after which both of them decrease with further increase in sintering temperature. Structural investigation shows formation of $Al_2O_3{\bullet}B_2O_3$ and $ZnO{\bullet}Al_2O_3$ at 900°C. However, $ZnO{\bullet}Al_2O_3$ disappears completely and $ZnO{\bullet}Al_2O_3$ becomes dominant. Further increase in sintering temperature leads to evaporation of glass, resulting in continuous decrease in density and strength. The process flow chart for SLS of Al_2O_3-B_2O_3 is given in Fig. 7.5.

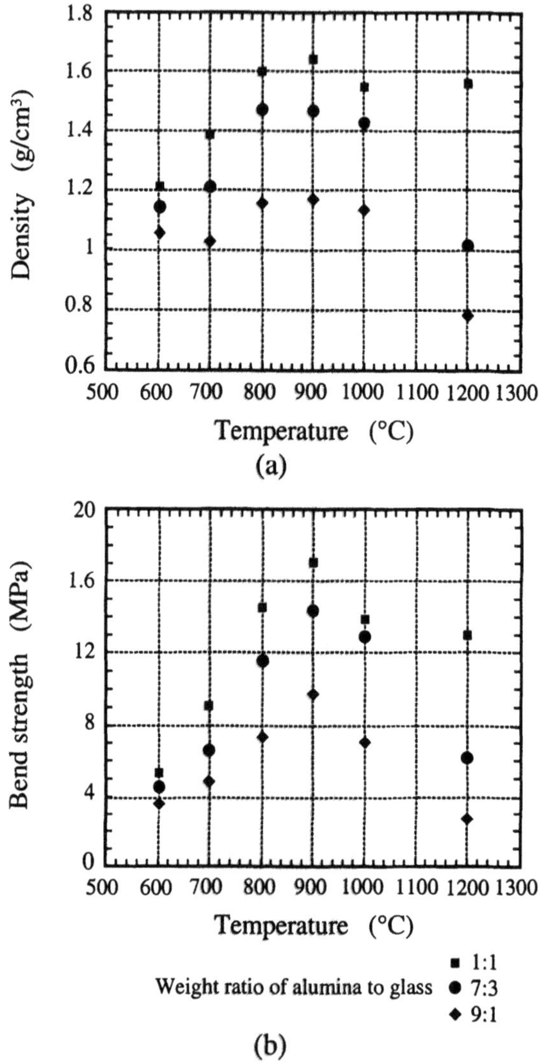

Fig. 7.4 Effect of amount of glass content and sintering temperature on density and bend strength of SLS alumina-glass composite [26].

As in the processing of alumina parts, zirconia parts have also been fabricated using SLS with infiltration. Zirconia powder for SLS is prepared by mixing and milling with copolymer. The SLS-processed parts need to be infiltrated by colloidal solution of zirconia. Wetting of the SLS processed parts can be improved by addition of a surfactant to the colloid [27]. A number of infiltration cycles may be performed to increase density of infiltrated parts. Debinding of the infiltrated part is needed by heating slowly to 500°C, followed by heating to 1500°C with a high ramp heating rate. If the furnace-sintered part is not dense enough, infiltration and sintering can be repeated [27]. Fig. 7.6 shows SLS and infiltrated parts with very fine micro-scale feature.

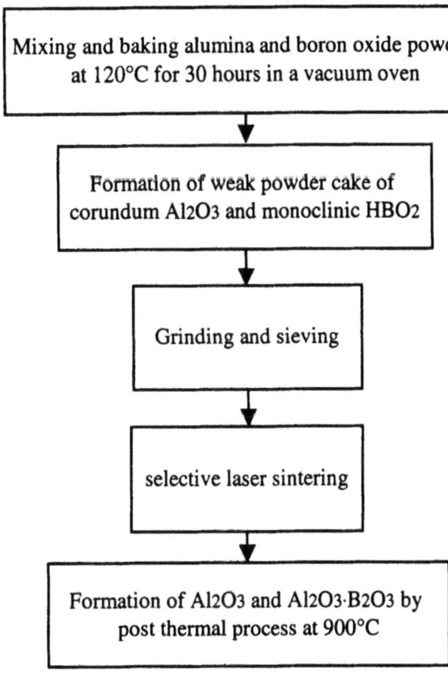

Fig. 7.5 Flow chart of SLS of alumina-boron oxide composite.

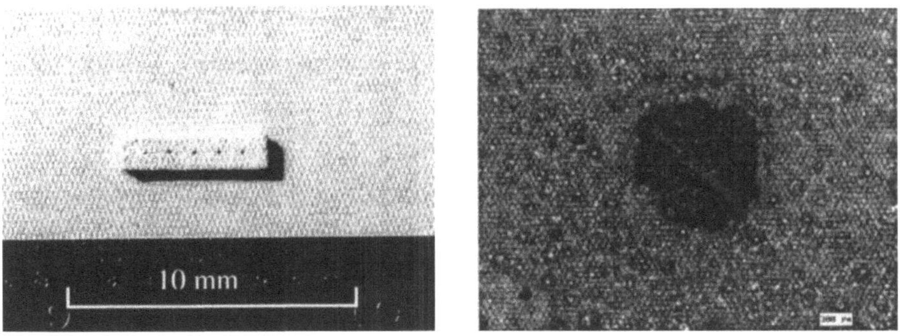

Fig. 7.6 A processed zirconia part with micron hole [27].

7.2.2 SLS with reactive binder

Another binder system is dihydro-Ammonium phosphate ($NH_4H_2PO_4$) that has been investigated by Lakshminarayan and Marcus [28]. This binder melts at 190°C and decomposes and reacts with Al_2O_3 at 850°C according to Bourell [17]:

$$Al_2O_3 + 2NH_4H_2PO_4 \rightarrow 2AlPO_4 + 2NH_3(gas) + 3H_2O(gas).$$

The final structure of this system is Al_2O_3-$AlPO_4$ composite with only less than 1% shrinkage. The advantage of using reactive binder is that the binder becomes part of the product through chemical reaction. Since no debinding is required, this process simplifies fabrication of ceramics. If the binder is less than 15vol.%, binding does not occur. Balling occurs when the binder exceeds about 60vol.% [17]. Compressive tests show increase in strength with binder volume fraction (Fig. 7.7). The increase in strength with binder content is partially attributed to the increase in density, and partially to the energy dissipated by the crack when it propagates through the material [29].

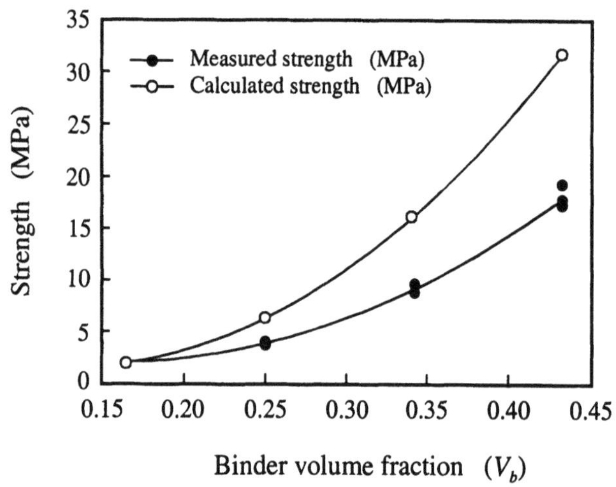

Fig. 7.7 Effect of binder on the strength of alumina-ammonium phosphate powder blends[29].

Bourell *et al.* [17] found that density can be increased using coarser (<100μm) Al_2O_3 compared to using fine one (<17μm). The improvement of density is attributed to an increasingly bimodal distribution of powder size in the mixture, which results in higher density of the powder bed before SLS.

7.2.3 SLS with infiltration reaction

In-situ ceramic formation shows some advantages over other processes [30-32]. SiC parts have been produced using this method where the SiC particles are first coated with polymer layer which acts as binder. SLS can then be carried out using low laser power energy. The processed part contains a large amount of open porosity that enables further infiltration to be performed. The first step of infiltration uses precursor resin in a vacuum system. Stierlen *et al.* [32] used an autoclave to cure the infiltrated parts under a pressure of 20 MPa at 170°C followed by pyrolysis and silicon infiltration similar to liquid phase silicon infiltration for ceramic matrix composites [33, 34]. During the second step, pyrolysis of precursor-infiltrated parts should be carried out at 900°C under a nitrogen environment. Carbon will be formed in this process. Due to carbonization, shrinkage takes place, leading to cracks in the carbonized structure. This cracked structure may be very helpful for the silicon infiltration. Formation of cracks are closely associated with the size and shape of the SiC particles, types of binder, infiltration resin and carbonization parameters.

The last step is silicon infiltration, which can be carried out at 1600°C in a vacuum condition to help remove gases from the porous SLS part. The infiltrated silicon reacts with the carbon in the prefom to form SiC. It is important to control the stoichimetric ratio of infiltrated silicon and residual carbon. Too little residual carbon may lead to incomplete reaction between Si and C. Therefore, the post-infiltration of precursor resin is crucial to maintain high percentage of carbon in the part. The full process chart is given in Fig. 7.8.

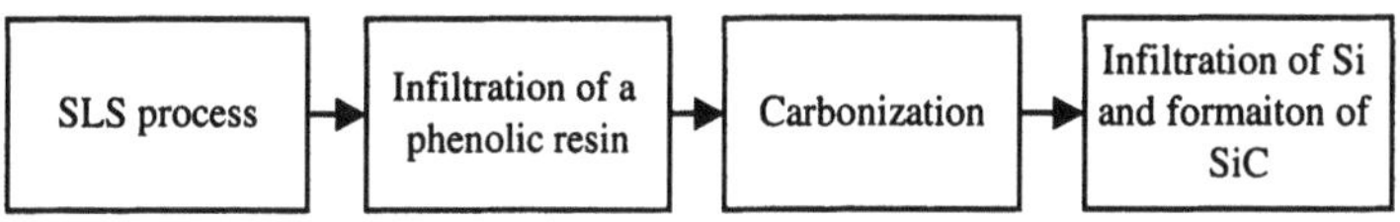

Fig. 7.8 Process chart of SLS of SiC part.

7.2.4 Selective laser reaction sintering

The main difference of selective laser reaction sintering (SLRS) from SLS is that reactions may take place during laser scan. Material systems used include ceramic-ceramic systems and particularly ceramic-metal systems.

One of examples of ceramic-metal system is SiC-Si in which SiC and Si is mixed before SLS process. SLS is carried out in a conditioned environment where methane gas is used. During laser scan, laser melts Si. At the same time, methane is decomposed releasing carbon. The Si is therefore truly carbonized in-situ forming SiC. No further post-process for the formation of SiC is needed.

In the SLRS, in general two reactants are required for the in-situ reaction. Table 7.1 lists some of powder systems for the SLRS process [39].

Table 7.1 Example of SLRS systems

System	Reactant 1	Melting point (°C)	Reactant 2	Melting point (°C)	Product
Al_2O_3-FeO	Al_2O_3	2070	FeO	1370	$Fe_xAl_{3-x}O_4$
Al_2O_3-Co_3O_4	Al_2O_3	2070	Co_3O_4	1795	$Co_x Al_{3-x}O_4$
Al_2O_3-NiO	Al_2O_3	2070	NiO	1985	$Ni_x Al_{3-x}O_4$
Al_2O_3-CuO	Al_2O_3	2070	CuO	1325	$Cu_x Al_{3-x}O_4$
Al_2O_3-MgO	MgO	2850	Al_2O_3	2070	$MgAl_2O_4$
Al_2O_3-SiO_2	Al_2O_3	2070	SiO_2	1710	$3Al_2O_3 \cdot 2SiO_2$
Y_2O_3-ZrO_2	ZrO_2	2700	Y_2O_3	2410	$Zr_{1-x}Y_xO_{0-0.5x}$

7.2.5 SLS of ceramics with metal binder

Due to the high melting temperature of ceramics, it is almost not feasible to directly melt ceramics using a laser. Hence, inorganic and organic binders have been developed for processing of ceramics by SLS. However, the inorganic binders may leave some constituent behind after furnace sintering, and careful control of debinding of organic polymer results in some difficulty. To overcome these problems and difficulty, metal binder has been developed [2, 35]. The interesting point of this process is the use of metal as a binder in the SLS process. The SLS-processed part is further oxidized during post-firing to form the same oxide as the matrix.

Subramanian *et al.* [27] has developed SLS of Al_2O_3 using Al as a binder. In their developed process, Al_2O_3 and Al powders are mixed. The Al_2O_3 powder particles are bound together into a part. After SLS, the part is fired to 800°C at 5°C/min. From 800°C to a temperature varying from 1000 to 1600°C, the heating rate is controlled at 2°C/min. During firing, residual Al is oxidized, depending on the firing temperature and duration. It is found that a large amount of Al is retained if the SLS-processed part is fired at 1200°C for a short time of 2 hours. Al is gradually reduced as the firing duration is prolonged. However, Al is still detectable even after 24 hours of firing. Al can only be oxidized after firing at 1600°C for 12 hours. It should be noted that full oxidation of Al to Al_2O_3 is not possible if a large part is fabricated. On the other hand, it is also not necessary to fully transform Al to Al_2O_3 since only the surface of the part is functionable.

It should be noted that wettability is an important consideration in the ceramic-metal system since wettability affects bonding between ceramics and metals during SLS. Table 7.2 lists wetting angles for a numbers of ceramic-metal system [36-38]. The contact angle depends on the temperature as well as environment. The wetting angle can be reduced if a vacuum system is used.

Al binder has also been used for SiC system [27] with 3:1 volume ratio of SiC to Al . Two kinds of powder are mixed. After forming SiC-Al part, it is fired at 1300°C in air for 18 hours. Oxidation of Al to Al_2O_3 takes place during firing. Complete oxidation of Al depends on the size of the part, firing temperature, and duration of firing.

7.2.6 SLS nano-sized powder

In general, the size of nano-powders is below 100 nm. Since the particle is so small, nano-powders offer very different properties from other larger-sized powders. Nano-powders have been used in SLS. Because different nano-sized powders are mixed, the diffusion length for solid-solid reaction is much shorter than that of micron-sized powders. The large surface area of nano-powders provides high driving force for fast sintering as well as reaction [39]. Therefore the sintering temperature can be lowered.

Table 7.2 Wetting angle for different ceramic-metal system

Materials system	Temperature (°C)	Condition	Contact angle
Al_2O_3-Al	700	-	120
Al_2O_3-Al	900	-	90
Al_2O_3-Al	100	-	75
Al_2O_3-Fe	1550	N_2	139
Al_2O_3-Ni	1450	Air	90
Al_2O_3-Cr	1950	Reducing	1-10
Al_2O_3-Cu	1200	Vacuum	138
Al_2O_3-Fe/Cr	1650	Reducing	~40
SiC-Al	725	-	150
SiC-Al	900	-	130
SiC-Al	1000	-	90
TiC-Fe	1550	H_2	39
TiC-Co	1500	H_2	36
TiC-Ni	1450	H_2	17
WC-Fe	1500	Vacuum	0
WC-Co	1500	Vacuum	0
WC-Ni	1450	Vacuum	0
WC-Cu	1450	Ar	30
TiC/WC-Co	1477	H_2	26
TiC/WC-Co	1477	Vacuum	4
TiC/WC-Ni	1477	H_2	16
TiC/WC-Ni	1477	Vacuum	6

7.3 Summary

SLS has shown some potential for use in processing of ceramics parts. Because of high melting points of ceramics, indirect SLS processes are used. Two kinds of binders are introduced, polymer-based and metallic-based. The former is decomposed during sintering process while the latter will react with other element to form in-situ ceramics. Due to the nature of the initial low laser sintering density, high shrinkage in the post process limits the use of SLS for ceramics processing. By introducing reaction process and infiltration, shrinkage can be reduced. The use of nano-sized powder enable lower sintering temperatures and also reduces the interdiffusion distance.

7.4 References

1. U. Lakshminarayan and H.L. Marcus, *Proc. Solid Freeform Fabr. Symp.*, TX, Ed. H.L. Marcus, J.J. Beaman, J.W. Barlow, D.L. Bourell and R.H. Crawdford, The University of Texas at Austin, Austin, 3-5 Aug. 1992, 44.
2. P.K. Subramanian, G. Zong and H.L. Marcus, *Proc. of the Solid Freeform Fabrication Symp.*, Ed. H.L. Marcus, J.J. Beaman, J.W. Barlow, D.L. Bourell and R.H. Crawdford, The University of Texas, Austin, 3-5 Aug. 1992, 63.
3. G.A. Brady and J.W. Halloran, *RP. J.*, Vol.3 (1997), 61.
4. D. Klosterman, R., Chartoff, N. Osborne, G. Graves, *Proc. Solid Freeform Fabr. Symp.*, Ed. D. L.Bourell, J. J. Beaman, R. H. Crawford, H. L. Marcus and J. W. Barlow, University of Texas at Austin, Austin, TX, 11-12 Aug., 1997, 537.
5. E.A. Griffin, D.R. Mumm, D.B. Marshall, *Amer. Ceram. Soc. Bull.*, Vol. 75, No. 7 (1996), 65.
6. M.K. Agarwala, V.R. Jamalabad, N.A. Langrana, A. Safari, P.J. Whalen and S.C. Danforth, *RP. J.*, Vol. 2 (1996), 4.
7. J. Yoo, M. Cima, E. Sachs, S. Suresh, *Ceram. Eng. Sci. Proc.*, Vol. 16, No.5, Sep-Oct, 1995, 755.
8. A. Safari, *J. Phys. III.* Vol. 4 (1994), 1129.
9. V. F. Janas, and A. Safari, *J. Am. Ceram. Soc.*, Vol. 78, No. 11, (1995), 2945.
10. W. A. Smith, *SPIE, New Developments in Ultrasonic Transducer and Transducer Systems*, Vol. 1733 (1992), 2.
11. K. A. Klicker, J. V. Biggers, and R. E. Newnham, *J. Am. Ceram. Soc.*, Vol. 64, No. 1 (1981), 5.
12. J.D. Cawley, A.H. Heuer, W.S. Newman and B.B. Mathewson, *Amer. Cer. Soc. Bull.*, Vol. 75, No.5 (1996), 75.
13. M. Greulich, M. Greul and T. Pintat, *RP. J.*, Vol. 1, No.1 (1995), 20.
14. K.J. Jakubenas, Y.L. Lee, M.S. Shaarawi, H.L. Marcus and J.M. Sanchez, *RP. J.*, Vol.3, No.2 (1997), 66.
15. H.L. Marcus, J.J. Beaman, J.W. Barlow and D.L. Bourell, *Am. Ceram. Soc. Bull.*, Vol. 69, No. 6 (1990), 16.
16. M.K. Agarwala, D.L. Bourell, A. Manthiram, B.R. Birmingham and H.L. Marcus, J. Mater. Sci., Vol. 30 (1995), 459.
17. D.L. Bourell, H.L. Marcus, J.W. Barlow and J.J. Beaman, *Intern. J. Powder Metall.*, Vol. 28, No. 4 (1992), 369.
18. P. K. Subramanian, N. Vail, J. Barlow and H. Marcus, *RP J.*, Vol. 1, No. 2 (1995), 24.
19. H.H. Bache, *J. Am. Ceram. Soc.*, Vol. 53, No. 12 (1970), 654.
20. A. Inaba, T. Kashiwagi and I.E. Brown, *Poly. Degradation Stability*, Vol. 21 (1988), 1.
21. T. Kelen, *Polymer Degradation*, Van Nostrand Reinhold Co., New York, NY, 1983.
22. N.K. Vail, J.W. Barlow, J.J. Beaman, H.L. Marcus and D.L. Bourell, *J. App. Poly. Sci.*, Vol. 52 (1994), 789.
23. N.K. Vail, *Preparation and Characterization of Microencapsulated, Finely Divided Ceramic Materials for Selective Laser Sintgering*, Ph.D Dissertation, The University of Texas at Austin, Austin, TX, 1994.
24. I. Lee, A. Manthiram and H.L. Marcus, *Proc. Solid Freeform Fabr. Symp.*, Ed. H.L. Marcus, J.J. Beaman, D.L. Bourell, J.W. Barlow and R.H. Crawford, The University of Texas at Austin, Austin, TX, 8-10 Aug. 1994, 339.
25. N.N. Greenwood and A. Earnshaw, *Chemistry of the elements*, Pergamon Press, Oxford, 1994.
26. I. Lee, *J. Mater. Sci. Lett.*, Vol. 17 (1998), 1907.
27. N. Harlan, S. M. Park, D. L. Bourell, J. J. Beaman, *Proc. of Solid Freeform Fabrication Symp.*, Ed. D. L.Bourell, J. J. Beaman, R. H. Crawford, H. L. Marcus and J. W. Barlow, The University of Texas at Austin, Austin, Texas, 9-11, 1999, 297.

28. U. Lakshminarayan and H.L. Marcus, *Proc. Solid Freeform Fabr. Symp.*, Ed. J.J. Beaman, H.L. Marcus, D.L. Bourell and J.W. Barlow, The University of Texas at Austin, Austin, TX, 12-14 Aug., 1991, 205.
29. U. Lakshminarayan and H.L. Marcus, *Mater. Manuf. Proc.* Vol. 9, No. 5 (1994), 921.
30. W. Löschau, Selektives Lasersintern von Strukturkeramik Jahresbericht FhG-IWS, Fraunhofer-Intitut für Werstoff und Strahltechnik, Dresden, 1997.
31. W. Löschau, R. Lenk et al, *Proc. 9th CIMTEC*, Florence, Italy, June 14 – 19, 1998.
32. P. Stierlen, P. Schanz and P. Eyerer, *Proc. Solid Freeform Fabric. Symp.*, Ed. D. L.Bourell, J. J. Beaman, R. H. Crawford, H. L. Marcus and J. W. Barlow, 10-12 Aug., 1998, 581.
33. P. Schanz and W. Krenkel, *Proc. 6th ECCM*, Bordeux, France, Sep. 20-24, 1993.
34. F. Gern, *Proc. 2nd HT-CMC*, Santa Barbara, California, Aug. 21-24, 1995.
35. P.K. Subramanian and H.L. Larcus, *Mater. Manuf. Proc.*, Vol. 10, No. 4 (1995), 689.
36. D.A. Weirauch, Jr., *Ceramic Microstructure' 86, Role of Interfaces*, Ed. J.A. Pask and A.G. Evans, 329.
37. C.A. Calow and A. Moor, *Practical Metallic Composites*, The Institute of Metallurgists, London, 1974, B17.
38. G.A. Shoales, *In Situ Strength Evolution in The Sintering of Bronze Powder Metal Compacts and Application to Thermal Cycle Design*, Ph.D. thesis, Pennsylvania state university , 1998.
39. A. Manthiram, D.L. Bourell and H.L. Marcus, *JOM*, Nov. 1993, 66.

Chapter 8

CHARACTERIZATION, MODELING AND OPTIMIZATION

8.1 Introduction

The successful fabrication of an RP part requires an understanding of the building process so that the most suitable RP material, machine, and processing considerations can be determined. There is a large variety of RP machines presently available in the market, catering to different requirements –Table 8.1 lists some of the commercially available RP machines. Most of the commercial RP machines can handle more than one kind of materials, with some providing four or five choices. Each RP system has its individual strengths and weaknesses: some can build models quickly; some can generate more precise or functional parts but at a relatively larger expense. The large variety of RP materials/machines and the strengths/weaknesses of individual RP processes enable different types of prototypes to be fabricated to meet specific performance and cost requirements. It also increases the complexities in the selection of a suitable combination of RP material, machine and building style for specific requirements. For example, the building style is constrained by the type of material and process/machine to be used. It concerns the building orientation for the part, the slicing of its three-dimensional (3D) model and the generation of support structure.

The building orientation of an RP part not only affects its quality, but also the time and cost. The effects of a part orientation on the manufacturing quality, time and cost differ in different RP processes. As a consequence, a suitable orientation needs to be selected to achieve desired goals.

The slicing of a 3D CAD model is an essential step for part fabrication with the layer-based RP techniques. The layer thickness of each slice not only affects the surface quality but also the manufacturing time of an RP part. Constant-layer thickness slicing is widely used in RP machines. With constant-thickness slicing, high surface quality demands the use of thin layers in the fabrication of the part, resulting in long fabrication time. Variable-thickness part building enables thin slices to be used in critical regions to achieve required surface roughness and thick slices in regions that do not, so that the overall fabrication time can be minimized without compromising required surface quality.

Table 8.1 Listing of RP Machines

Manufacturer	Country	RP machine
3D Systems	USA	SLA5000, SLA3500, SLA500/30, SLA500/20, SLA 500/10, SLA350/10, SLA 250/40, SLA 250/30, SLA 250/20, SLA 250/10, Actua 2100, ThermoJet
DMEC/Sony	Japan	JSC2000, JSC3000, SCS-1000HD
EOS GmbH	Israel	STEOS 400, STEOS 300, STEOS 600, STEOS 1200, EOSINT
CMET/Mitsubishi	Japan	SOUP 850, SOUP 530, SOUP 600, SOUP 400
Teijin-Seiki	Japan	Soliform 500, Soliform 300
Kira	Japan	Kira Solid Center KSC-50
Mitsui Engineering	Japan	COLAMM-300
Soligen	USA	DSPC300G, DSPC300
Cubital	USA	Solider 5600, Solider 4600
Stratasys	USA	FDM 8000, FDM 3000, FDM Quantum, FDM 2000, FDM 1650, 3D Modeler, Genisys Xs
Sanders	USA	ModelMaker II, ModelMaker III
Z Corporation	USA	Z402
DTM	USA	Sinterstation 2500, Sinterstation 2500 plus, Sinterstation 2000
LOM	USA	LOM 2030H, LOM 1015, LOM 1065
Kinergy	Singapore	Zippy I, Zippy II, Zippy III
Sparx AB	USA	Hot Plot
MIT/Soligen	USA	3D Printer
Schroff Development Corp.	USA	JP System 5
BPM Corporation	USA	BPM Personal Modeler
Delft Spline Systems	USA	DeskProto
Object Gedmetries	USA	Object Quardra

8.2 Modeling of RP part fabrication

In determining the appropriate RP process/material, it is important to first characterize and model the building mechanisms of different RP processes. Generic models applicable across several types of RP processes with similar characteristics can be developed for the evaluation of surface roughness, manufacturing time and cost of part/prototypes built.

8.2.1 RP processes

There are four relatively mature and popularly used processes among the variety of RP techniques: *laser lithography* (*LL*), *selective laser sintering* (*SLS*), *fused deposition method* (*FDM*) *and laminated object manufacturing* (*LOM*). Each of these four processes represents a distinct family of RP machines in the market. For examples, for the LL process, there are the SLA machine from *3D Systems*, the STEOS machine from EOS and the Solid Creation System from *SONY*. Although all four processes are based on the same layer building principle, they differ in the specific fabrication method.

The fabrication methods can be differentiated in terms of four key process parameters: the state of the raw material, the energy source, the layer creation process, and the condition of the completed part. The state of raw material concerns the physical state of the building material. The energy source refers to the energy type used to achieve the material addition in the building process. The layer creation process indicates the physical and/or chemical process to form a solid layer in the building process. The condition of the completed part describes the appearance and its surroundings. The characteristics of these four process parameters with respect to LL, SLS, FDM, and LOM are summarized in Table 8.2.

Table 8.2 Comparisons of four RP processes

	State of the raw material	Energy source	Layer creation method	Condition of the completed part
LL	Liquid	Laser beam	Additive: Photo curing	Immersed in liquid
SLS	Solid: powder	Laser/heater	Additive: Laser sintering	Buried in powder
FDM	Solid: filament	Heater	Additive: Fusion	Exposed to air
LOM	Solid: sheet	Laser/heater	Subtractive: Laser cutting	Buried in solid sheets

The type of building mechanism has a significant influence on the the building results, such as the dimensional accuracy, surface roughness, manufacturing time, and cost of parts built with different RP processes. For example, the state of the raw material is directly related to the surface roughness achievable with an RP process. Processes employing liquid raw material usually generate better surface roughness in RP parts than those using solid-state raw material. A negative effect with the liquid raw material is that the residual stress arising from the liquid-to-solid state transformation tends to lead to larger distortion in the built part.

The type of energy source used in the building process has an impact on the dimensional accuracy achievable with an RP process. In general, better dimensional accuracy can be obtained with processes employing a laser beam, due to the relative ease of controlling the laser beam. In FDM, the layers are formed with molten thermoplastic extruded from a nozzle. The attainable consistency and fineness of the width determine the achievable dimensional accuracy and surface roughness of an FDM-built part.

The manufacturing time and cost are closely related to the layer creation process and the condition of the completed part. In LOM, the raw sheet is first glued to the preceding sheet and the desired 2D profile is then cut out of the glued sheet. Large parts or models with generally solid interior can normally be built faster with such subtractive layer creation method. The conditions of the completed part determine the required post processing procedures. In LL, the completed part is immersed in liquid; and in FDM, the part is exposed to air. As the fluid medium in which the part is fabricated does not provide adequate support for some overhanging sections in the building process, specially constructed support structures may be needed to minimize distortion and inaccuracy of the built part. Depending on the geometric characteristics of the product model, the post processing procedures of the LL and FDM parts may involve appropriate finishing operations to remove the external supports. In SLS and LOM, the completed parts are buried in and supported by solid powder or stacks of sheets. No external supports need to be built to prop up overhanging sections of the part during the building process. However, the subsequent removal of the partially encased part can be time-consuming and difficult, depending on the geometric characteristics of the product model.

8.2.2 Surface roughness of a green part

The surface roughness of a green part is a function of several factors, including the state of the raw material, the layer creation process, the layer thickness, the building style and the resolution of the scanning/deposition system. Two kinds of surfaces least affected by the slicing layer thickness are the upward-facing and vertical planar surfaces. Instead, the surface roughness measurements from these surfaces reveal more of the effects of the building style and the building process. Appropriate benchmark parts can therefore be designed to determine such effects for different RP processes.

The surface roughness of the upward-facing horizontal and vertical planar surfaces in such benchmark parts built with the different RP processes have been measured with a Rank Taylor Hobson surface texture measuring and analysis system. This instrument adopts the ISO standard 4287-1:1984 (which is now replaced by ISO 4287:1997). The profile on the horizontal surface is measured across the scan line or (in FDM) material laying road direction in the fabrication of a sectional layer. The measurement on the vertical planar surface is taken along the successive layer addition direction. Four surface roughness profile parameters (R_a, R_t, R_{pm}, S) of the horizontal and vertical planar surfaces with the four specimens are listed in Tables 8.3 and 8.4, respectively. Parameter R_a is known as the arithmetic mean of the absolute ordinate values within a sampling length and R_t is the sum of the height of the largest profile peak height and the largest profile valley depth within the evaluation length. (Both R_a and R_t are still in use in ISO 4287:1997. R_{pm} is the mean of the roughness profile peak heights within a sampling length. In ISO 4287: 1997, R_p is used instead. Parameter S describes the roughness profile element width, which is represented by RS_m in ISO 4287:1997.)

Table 8.3 Surface roughness of the horizontal planar surface in the four RP parts

	R_a (μm)	R_t (μm)	R_{pm} (μm)	S (μm)
LL specimen	1.87	11.64	3.74	140.77
SLS specimen	16.27	104.79	39.74	85.76
FDM specimen	18.29	102.33	26.67	486.00
LOM specimen	9.62	142.78	28.58	66.63

Table 8.4 Surface roughness of the vertical planar surface in the four RP parts

	R_a (μm)	R_t (μm)	R_{pm} (μm)	S (μm)
LL specimen	3.54	21.94	10.21	150.84
SLS specimen	16.11	110.63	57.52	75.83
FDM specimen	18.75	70.86	31.76	123.44
LOM specimen	10.89	128.35	30.30	111.82

Tables 8.3 and 8.4 show that the values of R_a, R_t and R_{pm} in the LL specimen are all much smaller than those of the corresponding parameters in the other three RP parts. The plotted profiles also show that the surface roughness of the LL specimen is the best, giving the smallest magnitude of the deviation. The measured arithmetic mean deviation R_a of the upward-facing horizontal and vertical planar surfaces in the LL specimen is around 1.8μm and 3.5μm, respectively. Following LL is the LOM process that produces parts with R_a of around 10μm. The surface roughness of the SLS and FDM parts are much worse, with values of R_a around 16μm and 18μm, respectively. In the following section, the arithmetic mean deviation R_a is used in the analyses of the surface roughness caused by the staircase effect on an inclined planar surface of the RP part.

8.2.3 Surface roughness due to staircase effect

The surface roughness of an inclined planar surface built with layer-based RP technology is not only affected by the building style and the building mechanism, but also by the staircase effect, an effect common to all layer based RP technologies, as illustrated in Fig. 8.1(a). The extent of the staircase effect depends on the layer thickness and the inclination angle of the surface with respect to the building plane. Assuming constant layer thickness of L and an inclination angle α are used, as illustrated in Fig. 8.1(b), the mean line resulting from the staircase effect can be derived as:

$$\text{mean} = \frac{\{\int_0^{L\sin\alpha} x\cot\alpha\, dx + \int_0^{L\cos\alpha\cot\alpha} (L\cos\alpha\cot\alpha - x)\tan\alpha\, dx\}}{L/\sin\alpha}$$

$$= \frac{L\sin\alpha}{2\tan\alpha} \tag{8.1}$$

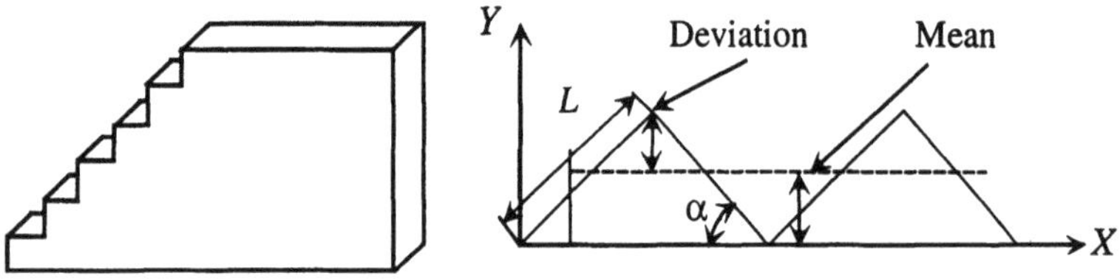

(a) Staircase effect in an RP Part (b) Evaluation of R_a

Fig. 8.1 Roughness caused by staircase effect in an RP part.

According to the definition of R_a [1], the roughness caused by the staircase effect can be expressed as:

$$R_a^s = \frac{1}{l_{cutoff}} \int_0^{l_{cutoff}} |y|\, dx$$

$$= \frac{1}{L/\sin\alpha} \left\{ \int_0^{L\sin\alpha} |x\cot\alpha - \text{mean}|\, dx + \int_0^{L\cos\alpha\cdot\cot\alpha} |L\cos\alpha - x\tan\alpha - \text{mean}|\, dx \right\}$$

where L_{cutoff} is the sampled length (= $L/\sin\alpha$). Replacing mean from Eq. (8.1),

$$R_a^s = \frac{L\sin\alpha}{4\tan\alpha} \tag{8.2}$$

In Fig. 8.1(b), both the horizontal and vertical profiles of a layer are expressed as straight lines. This is generally not encountered in the actual building process of RP techniques. Examples of the measured surface profiles of these two kinds of surfaces built with LL and FDM are shown in Fig. 8.2 and 8.3. From Fig. 8.2, it can be seen that the vertical surface profile displays some form of periodic characteristics. The maximum peak-valley deviation occurs between adjacent layers. The profile of the vertical planar surface can be seen as a result of two types of distortion: the distortion of the vertical profile in a single layer and the distortion resulting from the merging of layers. The first kind of distortion has a smaller period; it is likely to be filtered out in the processing of the surface roughness measurement. The computed R_a of the vertical planar surface can thus be regarded as the average magnitude of the second kind of surface distortion, which has a wavelength larger than the value of one single layer thickness.

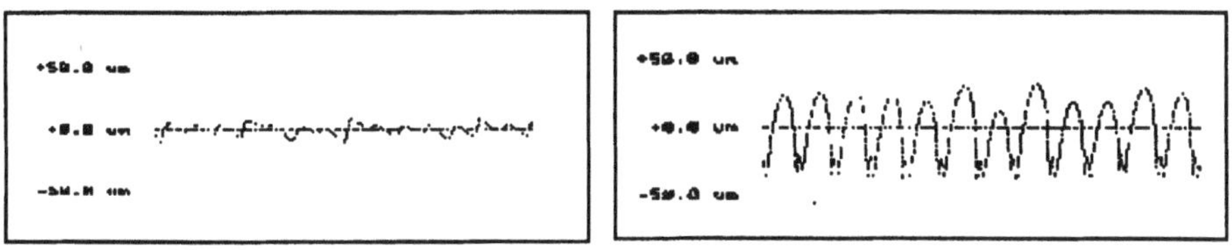

(a) LL specimen (b) FDM specimen

Fig. 8.2 Surface profiles of the vertical planar surface in the LL and FDM specimen.

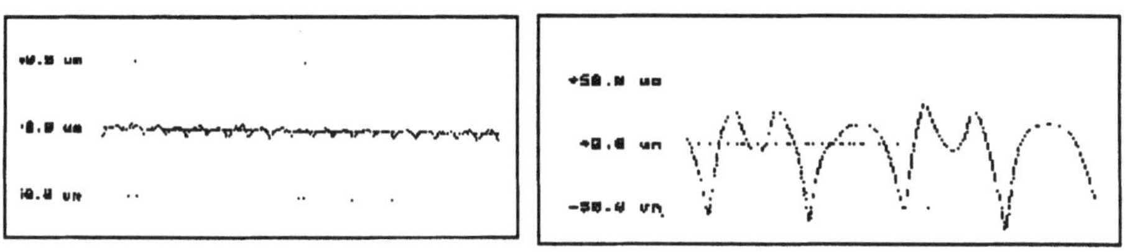

(a) LL specimen (b) FDM specimen

Fig. 8.3 Surface profiles of horizontal planar surface in the LL and FDM specimen.

The horizontal surface profiles also exhibit some form of periodic characteristics, as shown in Fig. 8.3. The periods of the dominant distortion patterns in the upward-facing horizontal surfaces built with LL and other laser-based RP processes are supposed to be relevant to the laser scanning pitch in the building plane. In the part built with FDM, the width of the fill road in the FDM machine, a process parameter set by the user, has a strong relation with the period of the dominant deviation pattern.

Since the dominant frequency of the deviations resulting from the vertical building inaccuracy and the horizontal building profile are not less than one single layer thickness in most cases, the effects of these two kinds of deviation patterns on the roughness of the inclined planar surface cannot be ignored. Thus, a way to predict the surface roughness of an inclined planar surface in an RP part is to combine the effects from the aforementioned three sources: the staircase effect, the horizontal deviation pattern and the vertical deviation pattern. One approach is to use the surface roughness of the upward-facing planar surface R_a^h to represent the average deviation of the horizontal building profile and use the surface roughness of the vertical planar surface R_a^v to represent the deviation of the vertical building profile, as illustrated in Fig. 8.4. As the deviations of the vertical and horizontal profiles are not perpendicular to the mean line, the projections of these two factors are considered. The roughness contributed by the deviation patterns along the horizontal and vertical directions can thus be simplified as:

$$\left(\int_0^{L\sin\alpha} R_a^v \sin\alpha\, dx + \int_0^{L\cos\alpha\cdot\cot\alpha} R_a^h \cos\alpha\, dx\right)/(L/\sin\alpha)$$
$$= R_a^v \sin^3\alpha + R_a^h \cos^3\alpha \qquad (8.3)$$

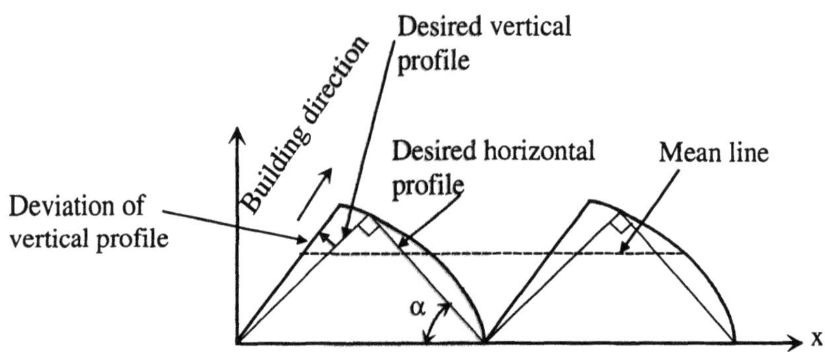

Fig. 8.4. Surface roughness of inclining planar surfaces in an RP part

Combining Eq. (8.2) and (8.3), the following can be derived:

(i) $R_a = R_a^h$, when $\alpha = 0°$;

(ii) $R_a = R_a^v$, when $\alpha = 90°$;

(iii) $$R_a = \frac{L\sin\alpha}{4\tan\alpha} + R_a^v \sin^3\alpha + R_a^h \cos^3\alpha, \quad 0<\alpha<90 \qquad (8.4)$$

where L is the layer thickness, and α is the acute angle between the normal vector of the referenced surface and the building direction. R_a^h and R_a^v can be found by experimental study.

The prediction based on Eq. (8.4) is consistent with the experimental result presented by Reeves and Cobb [2]. A study of the surface fault patterns in parts built with SLS also demonstrates that the surface fault pattern is primarily due to the staircase effect [3]. The secondary surface fault patterns relate to the finite size of the powder, the vibration of the roller and the width of the scanning line. The effects resulting from these three factors dominate the surface fault pattern of the up-facing planar surface in the SLS specimen. The third surface fault pattern is related to the layer shifting, which can be seen clearly in the vertical planar surface in the SLS specimen.

The surface roughness of the planar surface in an RP part can be evaluated based on Eq. (8.4). Usually the surface roughness parameters R_a^h and R_a^v depend on the material type, the building style and the process parameter settings with a certain RP machine. When these factors are constant, the achievable surface roughness of a given planar surface in a green RP part can be estimated before the actual fabrication.

8.2.4 Part fabrication time

The manufacturing time of an RP part is defined as the time between submission of the design and the completion of the finished prototype/part. The time to manufacture an RP part can thus be separated into three parts: data preparation time, fabrication time and post processing time.

Data Preparation Time

The data preparation stage is similar in all layer-based RP techniques. It involves file transfer, data conversion, orientation selection and model slicing, layered data sorting and merge, process parameter setting and control file generation. For processes where external supports are needed to build the overhang, additional time will be required for the design and slicing of the support structures. Most of these data preparation tasks can be software-assisted. As the portion of data preparation time is small relative to the entire manufacturing period, the total data preparation time with a specific RP process can be considered to be a constant value.

Fabrication Time

For most layer-based RP processes, the fabrication time can be seen as the sum of the solidifying time for the solid volume and the delay time between the formation of layers. Eq. (8.5) can be used to calculate the fabrication time with most layer-based RP processes, such as LL and SLS.

$$T_f = nt_w + (V_{part} / L) / r_f + (V_{spt} / L) / r_{spt} \tag{8.5}$$

where T_f denotes the total fabrication time; n denotes the number of layers; L denotes the layer thickness; V_{part} denotes the volume of the product model;

V_{part} / L indicates the total section area needed to be solidified; t_w is the delay between layers; r_f is the solidifying rate for part geometry in XY plane; r_{spt} is the solidifying rate for support structure in XY plane; V_{spt} is the volume of external support.

Factors such as r_f, r_{spt} and t_w are all process-related. They can be computed based on the process parameter settings within each RP machine. Because of its special process characteristics, the delay time in the LL process has the greatest influence, consisting of a number of components, as shown in Table 8.5.

Table 8.5 Idle time between layers in three RP processes

	Delay time t_w between the formation of layers
LL	Coating time, pause time, pip delay time, platform elevating time
SLS	Coating time, platform elevating time
FDM	Nozzle elevating time

The computation of the fabrication time with LOM, however, is slightly different. The layers of a LOM part are cut out of sheets of paper, plastic or composite by the laser. The 3D solid part is essentially built within a block of sheets so that the part can be protected during the building process. Excess material beyond the boundary of the part is cut into small pieces or crosshatched during the fabrication of each layer. Hence, the fabrication time of a layer consists of two portions: contour cutting time and crosshatching time. The crosshatching time can be estimated based on the laser scanning speed and the crosshatch pitch. The estimation of the contour cutting time is more difficult. One approach is to find the total length of the contour through the simulation of the slicing process, and then divide the total length by the laser scanning speed to get the total cutting time. Another is to roughly estimate the contour cutting time as follows: the building part is simplified as a rectangular block with known volume and height; the contour cutting time can be computed using Eq. (8.6).

$$T_{scan} = n(V_{part} / Z_{max})^{1/2} 4 / v_l \qquad (8.6)$$

Since the contour cutting time is much less than the cross-hatching time for most cases, the fabrication time of a LOM part can thus be estimated using Eq. (8.7).

$$T_f = nt_w + \frac{2X_{max} Y_{max}}{v_l \, l_{pitch}} n + T_{scan} \qquad (8.7)$$

where X_{max}, Y_{max}, and Z_{max} represent the building envelope of the part; v_l denotes the scanning speed of laser; l_{pitch} represents the crosshatch pitch; n is the number of layers; V_{part} is the volume of the building part.

Given a building direction, the length of the building model in the X or Y direction, i.e. the envelope of model in the XY plane, can be found by projecting all surfaces onto the XY plane and then computing the union of the projections.

Table 8.6 Post processing tasks in the four RP processes

	Post Processing Steps
LL	Removing liquid resin, removing support if support exists, post curing, sand finishing (optional)
SLS	Removing powder, polymer infiltration (optional), drying (optional), sand finishing (optional)
FDM	Removing support if support exists, sand finishing (optional)
LOM	Removing redundant material, sand finishing (optional)

Table 8.7 Classification of post processing tasks in the four RP processes

	LL	SLS	FDM	LOM
Cleaning	Yes	Yes	Geometry dependent	Yes
Post curing	Yes	No	No	No
Finishing	Yes	Yes	Yes	Yes

Post Processing Time

Prototypes built with almost all RP processes will require some form of post processing. Table 8.6 lists all major post processing steps with the four RP processes. For simplification, the post-processing tasks can be grouped into three stages: cleaning, post curing and finishing. The post processing time, which is the summation of time spent in the three stages, depends on the tools used for processing and the geometric complexity of the part. Table 8.7 shows the post-processing requirements of the four RP processes in three stages.

8.2.5 Part fabrication cost

The manufacturing cost of an RP product/prototype can be computed based on following components:

- Material cost C_{mat}.

- RP machine running cost. This includes the depreciation/usage of machine, the electricity, maintenance and other overhead in the part fabrication stage.
- Cost in data preparation stage. This includes the computational cost related to the use of hardware/software and the labour cost in the data preparation stage.
- Cost in the post processing stage, including labour cost and overhead.

Material Cost

The material cost to build an RP part varies with the RP process. The difference not only comes from the different price of the material, but also on the part geometry and the varying characteristics of the RP process. For processes requiring external supports, such as LL and FDM, the material cost can be divided into two components: the cost for the part building and the cost for building the support structure.

$$C_{mat} = P_{m\_p} V_{part} + P_{m\_s} V_{spt} \tag{8.8}$$

where P_{m_p} is the unit price of part material; P_{m_s} is the unit price of support material; V_{part}, V_{spt} are volumes of the part and external support, respectively.

In LL, the same material is used for both part building and support fabrication and hence, P_{m_p} equals P_{m_s}. Due to the differences in the building mechanism, the volume of external support needed for part fabrication with the LL and FDM could be different. Evaluations of the volume of support structures in LL and FDM have been reported previously [3,4].

For RP processes with inherent support, such as SLS and LOM, the material cost relies on the material type, the layer creation method and the re-usability of the raw material. For example, SLS does not consume excess material during the creation of each layer. The surrounding powder that provides support for part during the building process can be reused in later runs. Hence, the material consumption in SLS only accounts for the buildup of the part geometry. In the case of LOM, the situation is more unique. The excess material surrounding the part cannot be directly reused without further processing. The total material consumption in LOM part fabrication thus should be computed based on the envelope of the part geometry and the thickness of the surrounding wall. Eq. (8.9) and (8.10) can be used to compute the material cost in SLS and LOM part fabrication, respectively:

$$C_{mat}^{sls} = P_{m\_p}^{sls} V_{part} \tag{8.9}$$

$$C_{mat}^{lom} = P_{m\_p}^{lom} (X_{\max} + \gamma_{wall})(Y_{\max} + \gamma_{wall})(Z_{\max} + \gamma_{bt}) \tag{8.10}$$

where $P_{m\_p}^{sls}$, $P_{m\_p}^{lom}$ refers to the unit cost of the SLS and LOM material; X_{max}, Y_{max}, and Z_{max} is the building envelope of the part; γ_{wall} denotes the thickness of surrounding wall; γ_{bt} is the thickness of the supporting layers at the bottom.

Machine Running Cost

The machine running cost for a given part can be evaluated based on the machine running rate and the fabrication time. Evaluation formulas for the fabrication time with different RP processes have been discussed earlier. The machine running rate reflects the capital cost of a RP machine, the power consumption, the labour cost, the maintenance cost and other overheads in the machine running stage. Assuming the hourly salary of the operator is w_0, the overhead for operator is o_{op}, the original cost of machine is P_{mch}, the overhead for machine running is o_{mch}, the amortization period of the machine is T_{mch}, the machine running rate can be computed using Eq. (8.11).

$$r_f = (1+o_{op})w_0 + (1+o_{mch})\frac{P_{mch}}{8760T_{mch}} \tag{8.11}$$

Eq. (8.11) is similar to the formula for computing the manufacturing cost with machining [5]. Suppose the hourly salary w_0 is \$10, the operator overhead o_{op} is 100%, the machine overhead o_{mch} is 200% and the amortization period of an RP machine T_{mch} is 3 years (which is commonly used in machining technology), the computed running rates with four RP machines are as shown in the third column in Table 8.8. Some RP service providers, especially those government-supported institutions, do not charge the operator's wage to the customer. In this case, the hourly rate can be cheap. The fourth column in Table 8.8 gives the computed machining running rates when the additional manpower cost is ignored, the machine overhead is 50% and the amortization period is 5 years.

Given the machine running rate, the running cost for a specific part can be computed based on the estimated fabrication time.

$$C_f = r_f T_f \tag{8.12}$$

Cost in the Data Preparation Stage

The cost in the data preparation stage consists of two items for all RP processes: the computation cost and the labor cost. Both items can be evaluated based on the time needed in the preparation stage. In Eq. (8.13), w_0 is the hourly salary of the operator, and w_{cp} is the computation cost rate. Both values are assumed to be the same for all RP processes. Hence, the difference of cost in the data preparation stage for each RP process is proportional to the preparation time T_{pre}.

$$C_{pre} = (w_0 + w_{cp})T_{pre} \tag{8.13}$$

Table 8.8 Computed machine running rate of the four RP machines

RP machine	Price ($)	Hourly rate I	Hourly rate II
JSC 2000 (SCS)	500,000	77 $/h	17.12 $/h
DTM 2000 (SLS)	300,000	54 $/h	10.27 $/h
LOM 1015 (LOM)	92,600	30 $/h	3.17 $/h
Concept Modeler (FDM)	55,000	26 $/h	1.88 $/h

Cost in the Post Processing Stage

Cost in the post processing stage is related to the complexity of the post processing procedures, the equipment cost, the electricity consumed and the labour cost. As mentioned earlier, the post processing work of an RP part can be organized into three tasks: cleaning, post curing and finishing. The total post processing time is the combination of the time spent on the three tasks. The cost in the post processing stage can thus be evaluated based on the time spent on each post processing task and the hourly charge rate for each task. As the equipment involved is different, the hourly charge rate differs in the three tasks. The hourly wage of the operator, on the other hand, can be assumed to be constant. The cost in the post processing stage can thus be computed using Eq. (8.14).

$$C_{post} = T_{cleaning}\gamma_1 + T_{post\_cure}\gamma_2 + T_{polishing}\gamma_3 + T_{post}w_0 \tag{8.14}$$
$$T_{post} = T_{cleaning} + T_{post\_cure} + T_{polishing}$$

where w_0 is the operator's hourly wage, γ_1, γ_2, and γ_3 are the hourly charge rates denoting the resource used in the cleaning, post curing and finishing tasks, respectively.

The computation of manufacturing time and cost in RP part fabrication entails the determination of process parameter settings in the RP machine, the machine/material prices and the geometric information of the product model. The geometric information involved includes the building envelope of the part, the volume of the part, the volume of external support structure, the slicing layer thickness, etc. The preceding equations are for the evaluation of surface roughness in a green RP part and the estimation of manufacturing time/cost, both of which are essential to the implementation of a decision support system for part fabrication with RP.

8.3 Optimal orientation

The aim in determining the optimal orientation for part fabrication with RP is to minimize the effect of the building orientation on the surface roughness and geometric features of the part for a particular RP process. The dissimilar building mechanism of each RP process affects the quality of the fabricated part in varying degree according to its orientation. The major sources of geometric inaccuracy affected by part orientation are the staircase effect, over-curing effect under the overhang area, part distortion due to instability of the part being built, and increased surface roughness caused by the removal of the external support structure attached to the under side of the overhang section to prop it up. Among these, some are common to all layer-based RP processes, e.g. the staircase effect. Some are associated to specific processes.

By considering important orientation-related sources of geometric errors mentioned above, various algorithms and software tools have been developed to identify the most suitable orientation in the fabrication of an RP part. Optimal orientation algorithms have been proposed based on: the minimization of the support contact area [6]; the number of slices and the ratio of staircase area versus the total surface area [7]; the desired part quality, fabrication time and volume of external support structure for LL part fabrication [8]; the part accuracy and fabrication time in SLA as the primary and secondary objectives, respectively [9]; a mathematical model that accounts for the SLA-built surface quality, height of the part, and support requirements from the perspective of the part geometry[10]. A rule-based expert system has also been developed for the selection of optimal build-up orientation with SLA[11]. The rules are defined based on the consideration of surface roughness, fabrication time and support structure. The rule-based expert system allows the user to consider special features and brings flexibility into the orientation selection process.

Most of the aforementioned approaches focus on the selection of optimal orientation for part fabrication with SLA, while a few concern the orientation for FDM. The following sections describe general considerations for determining orientation candidates for RP part fabrication and present examples of single- and multi-criterion search for the most appropriate orientation to build a given RP part.

8.3.1 Orientation candidates for RP part fabrication

The selection of part orientation with RP processes involve considerations of the following four factors:

- the stability of the part during fabrication such that it will be unlikely to topple over;

- the overhang area, which is related to the over-curing effect in LL and the external support contact area in LL and FDM;
- the number and area of inclined surfaces, which determine the volume of stair-case effect in layer-based RP processes;
- the part building height, which could be a measure of the fabrication time in the layer-based RP process.

Some RP processes, such as LOM and SLS where the part is buried in laminates or powder, provide inherent support to the part during the building process. In such processes, the stability of the part and the extent of surface overhang are not major concerns. As a consequence, it is the other two factors - the staircase effect and the building height – that need to be considered in the selection of a suitable orientation.

8.3.2 Single-criterion optimal orientation

The volume of staircase effect and the part building height are two major concerns in the selection of part orientation with most RP processes. The staircase effect is a common error source affecting the surface roughness of an RP part in the layer-based RP processes, while the part building height largely determines the manufacturing time in many RP processes. Hence, in obtaining the optimal orientation based on a single criterion, the minimization of one of these two kinds of features is usually considered.

Optimal Orientation with Minimum Staircase Effect

Given a constant layer thickness, the staircase effect in an RP part can be represented by the number or area of the inclined surfaces in the referenced building direction. It can be shown that the search for the optimal orientation with minimum staircase effect is limited to a finite set of orientations [12].

The staircase volume associated with an inclined planar surface is related to the surface area and the inclination angle of the surface with respect to the building plane, as illustrated in Fig. 8.5. Suppose the area of the planar surface is A_i, the layer thickness is L, the intersecting angle between the normal vector of the plane and the building directional vector is θ. The staircase volume of the inclined planar surface can be evaluated based on the summation of the individual step volumes V_{step} according to Eq. (8.15).

$$V_{step} \approx A_i \, L \cos\theta / 2 \tag{8.15}$$

With Eq. (8.15), the computed staircase volume can be positive (when θ is less than 90°), or negative (when θ is between 90° and 180°). As the concern is the extent of the staircase effect, the absolute value is used. Currently in RP, polyhedrons (part models bounded by tessellated surfaces) are commonly used in the representation of a 3D solid model. For a polyhedron, the search for the optimal orientation with the minimum staircase effect can proceed as follows:

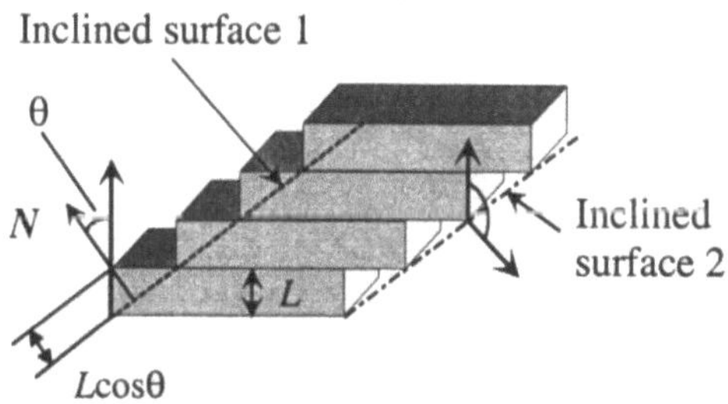

Fig. 8.5 Staircase volume with the inclined planar surface.

(1) Traverse the solid polyhedron and collect the normal vector of every planar surface into a set ***N***. As each new normal vector is collected, it is compared with those in ***N***. If the new normal vector is parallel with one of the of the vectors already in ***N***, it is dropped; otherwise, it is taken into the set ***N***.

(2) Compute the cross product for every pair of normal vectors in ***N*** and store the results in a set ***N***.

(3) Compare each vector in ***N*** with those in ***N*** by computing the dot products of two vectors. Remove those that are parallel vectors in both sets and then combine the two into a single set, say Ω.

(4) For each vector in Ω, its reverse direction is obtained and added to Ω, if it is not already a member.

(5) Taking each vector in Ω as the building direction, traverse the surfaces in the polyhedron, find the surfaces that are inclined and compute the staircase volume of each inclined surface in the orientation using Eq. (8.15). The sum of the staircase volumes in each orientation is then determined.

(6) The orientation that has the least staircase volume is identified and chosen as the optimal orientation.

Optimal Orientation with Minimum Building Height

Given a constant layer thickness, the part building height is proportional to the number of layers required for the completion of the part. In some RP processes, e.g., LL, the delay time between the fabrications of consecutive layers constitutes a large portion of the entire fabrication time. In such processes, identifying an optimal orientation that minimizes the building height can reduce the fabrication time significantly. Meanwhile, in certain processes such as LOM, the part building height is directly related to the volume of material consumed in the process. The minimization of the part building height with LOM will therefore also reduce the amount of material consumed, and decrease the overall manufacturing cost.

The search for the optimal orientation with minimum part building height proceeds in following steps:

(1) Construct the convex hull of the polyhedron.

(2) Use a face on the convex hull as the building base. For each vertex on the convex hull, the distance between the vertex and the base face can be computed. The maximum value is regarded as the part height in the referenced orientation.

(3) Compute the cross product of every pair of non-adjacent and non-parallel edges on the convex hull. Use each of these vectors as the building directional vector and compute the distance between the pair of edges, which gives the part building height in that orientation.

(4) Compare the part building heights in different orientations. The direction with the minimum building height is taken as the optimal orientation.

Multi-criterion Optimal Orientation

Optimal orientation based on the consideration of multiple criteria is of more practical use for decision-making pertaining to part orientation in RP. In multi-criterion consideration, factors affecting inaccuracy and manufacturing time are first identified and quantified. Based on the quantification of the manufacturing factors and the specific process constraints, a multi-criterion optimal orientation algorithm is then developed. In the following, a multi-criterion consideration for determining a suitable orientation for LL-built part is presented.

8.3.2.1 Quantification of building inaccuracy

Building inaccuracy is defined here as the volume of shape distortions affected by part orientation in RP. Sources of building inaccuracy with different RP processes are not completely the same. Table 8.9 lists the major sources of building inaccuracy found in the four RP processes. Geometric features related to the four kinds of building inaccuracy are listed in Table

8.10. Based on the inaccuracy sources listed in Tables 8.9 and 8.10, a quantitative estimate can be formulated in the form of building inaccuracy volume affected by part orientation in the four listed RP processes. For examples, in LL, the volume of building inaccuracy source can be quantified from the sum of the staircase volume, the oversize in the overhang area and the extent of distortion caused by the removal of external supports in overhang regions. In SLS, the volume of building inaccuracy can be defined as the sum of the staircase volume and the oversize in the overhang regions. (Oversize refers to extra material accumulating under the overhang part. This is due to curing of the photo-polymer liquid immediately below the overhang as a result of penetration by the laser through the successive layers above the overhang during the building process.)

Table 8.9 Inaccuracy sources related to part orientation in the RP processes

	Staircase	Z*-Oversize	Removal of external Support	Surface curling
LL	Yes	Yes	Yes	Yes
SLS	Yes	Yes	No	Yes
FDM	Yes	No	Yes	Yes
LOM	Yes	No	No	No

****Building direction***

Table 8.10 Geometric features relevant to the inaccuracy sources

Geometric features	Inaccuracy sources
Inclined surfaces	Staircase effect
Overhang	Oversize in Z direction due to over cure
Overhang	Removal of external support
Large flat surface	Curling

It can be seen from Table 8.10 that the inclined surface and the overhang on a surface are the two most important geometric features affecting the manufacturing quality of an RP part. The following sections show algorithms for the estimation of the amount of these two features in a solid model.

Computation of Overhang Region in a Solid Model

The inner product of the normal vector and the building direction vector can be used to determine if a planar surface is overhanging in a particular orientation. When the sign of the inner product is negative, the surface overhangs with respect to the building direction. The extent of surface overhang can be evaluated based on the projection of the surface area on the building plane. Suppose the planar area is A, its normal vector is N and the building direction vector is γ, the extent of surface overhang can be evaluated using Eq. (8.16).

$$A_{overhang} = A \mid N \cdot \gamma \mid \quad (8.16)$$

In a curved surface, the normal to the surface changes from point to point. Theoretically the overhang area on the curved surface can be found by evaluating the inner product of the normal vector with the building directional vector at different points on the surface, as shown in Fig. 8.6. The computation of the overhang area (*OHA*) in a surface involves a two-dimensional integral:

$$OHA = \iint_{u, v \in \psi} \left| \boldsymbol{r}_u \times \boldsymbol{r}_v \right| du dv \quad (8.17)$$

where Ψ is a subset in the parametric space addressed by the coordinates (u, v). Members in Ψ satisfy the equation $N(u, v) \cdot \gamma < 0$.

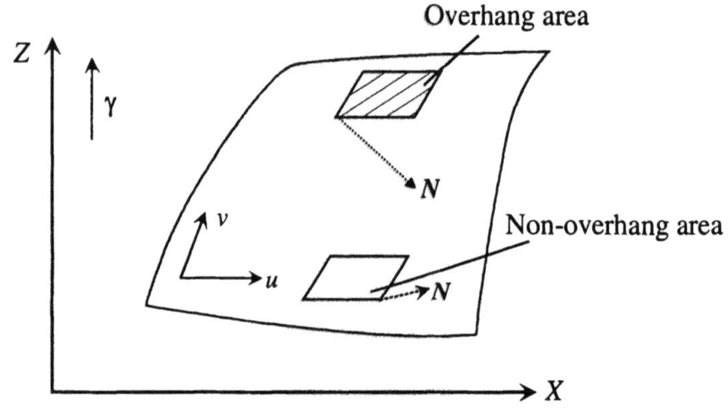

Fig. 8.6 Estimation of overhang area in a sculptured surface.

Given an analytic equation, the subset Ψ in the surface can be found by analytical means and the integral may be obtained analytically. A more general approach is to tessellate the curved surface into facets. The overhang region in a curved surface is then estimated based on the summation of overhang regions of those facets. Some form of sampling can be applied in the search for the overhang regions from the tessellated facets. A sampling method to estimate the overhang area in a curved surface is presented Fig. 8.7.

Sampling methodology has been used in reverse engineering of curved surfaces, surface inspection and surface data fitting [13]. By using a sampling strategy, a continuous curved surface is digitised into a set of discrete points. Intuitively the sample size is in proportion to the approximation accuracy of the surface. It has been shown that there exists a lower bound in the discrepancy for a finite sampling size [14]. The Hammersley sequence, one of the few sequences that have been reported to achieve near optimal low discrepancy [13, 15], is used as the sampling strategy in this algorithm.

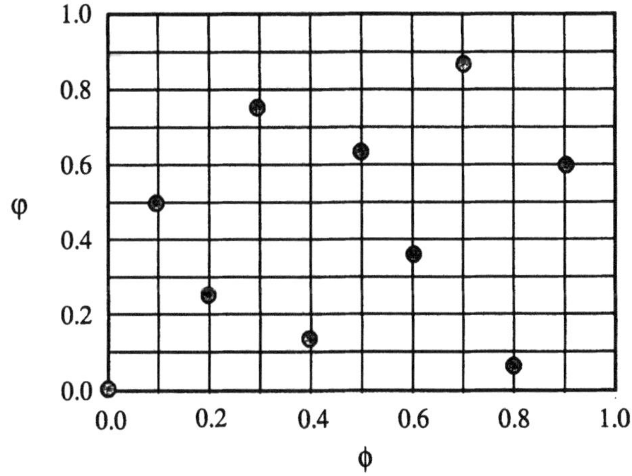

Fig. 8.7 Distribution of the ten-point Hammersley Sequence

Assume that the parametric space of the curved surface is a unit square addressable by parameters (ϕ, φ), where ϕ , $\varphi \in [0,1]$. In the parametric space, the corresponding points from a 2D Hammersley sequence are expressed by:

$$\phi_i = i / n$$
$$\varphi_i = \sum_{j=0}^{l} b_j 2^{-j-l} \tag{8.18}$$

where $i \in [\,0, n\text{-}1]$, l = Upper (log(i)), b_j is the jth binary digit of number i.

If n =10, then l = log$_2$(9) = 4, the sampled points in the square addressed by (ϕ, φ) would be:

ϕ_i = 0/10, 1/10, 2/10, 3/10, 4/10, 5/10, 6/10, 7/10, 8/10, 9/10; and
φ_i = 0/16, ½, ¼, ¾, 1/8, 5/8, 3/8, 7/8, 1/16, 9/15.

The distribution of the ten-point Hammersley sequence in a 2D space is shown in Fig. 8.7.

The estimation of the overhang area in a sculptured surface of area A_i proceeds as follows:

(1) *Initial Search.* Take $n = n_1$. Sample n_1 points using the set of parameters (ϕ_i, φ_i) computed from Eq. (8.18) on the surface and find the normal at each point.

(2) Compute the dot product of the normal vector with the building directional vector. If none of the dot products is negative, the overhang region on the surface is assumed to be zero. If a negative result is found, go to step 3.

(3) *Refined Search.* Take $n = n_2$, usually $n_2 >> n_1$. Sample n_2 points on the surface using a new Hammersley sequence with $n = n_2$. Each point is checked to see if it is in the domain of the surface (such as by using a suitable function provided by the CAD software. Points, denoted as *n_pt,* lying in the surface are counted and the normal vector is found at each such point;

(4) Determine the inner product of the normal vector with the building directional vector. The number of points, denoted as *neg_pt*, where the normal vector has a negative projection in the building direction is counted;

(5) The overhang area *OHA* is estimated using following equation:

$$OHA = \frac{neg\_pt}{n\_pt} A_i \tag{8.19}$$

The two-step sampling strategy is used in the algorithm to improve computation speed. The initial search enables overhang regions to be located quickly. The larger sampling size in the succeeding search is then used to obtain a more accurate estimation of the overhang region.

Evaluation of the Inclined Surface Area in a Solid Model

The volume of staircase effect in an RP part is determined by the inclination angle of the surface, the surface area, and the layer thickness used to build the part, as reflected in Eq. (8.15). Suppose a constant layer thickness is used, the inclination angle and the inclined surface area will then determine the extent of staircase effect in the specific orientation.

8.3.2.2 Part stability during the building process

The stability of the part during the building process is important for those RP processes that inherently, do not provide any support to the part as it is built, such as the FDM and LL processes.

When considering the part stability in the search for optimal orientation, a penalty function $F(\gamma)$ can be defined. $F(\gamma)$ is proportional to the projected distance c of the CG from the convex hull of the building base, the part height h in the specific orientation, and inversely proportional to the area of the building base A_{base}, as given in Eq. (8.20).

$$F(\gamma) = c\frac{h}{A_{base}} \tag{8.20}$$

The convex hull can be obtained over a set of distinct points, such as those sampled from the part surface. One simple method is to use the upper-and-lower chain algorithm [16, 17].

8.3.2.3 Effect of part orientation on manufacturing time

Part orientation primarily affects the fabrication time, i.e. the time to build the green part. Its effect on the other components of the manufacturing time, i.e. the preparation and post processing times, is relatively small. Hence, it is generally adequate for the optimal orientation algorithm to consider the minimization of the fabrication time only. For all layer-based RP processes, the fabrication time consists of two parts: the delay time between layers and the material formation time of each layer, which includes the layer for the formation of the support structure, as given by Eq. (8.5) in the earlier section. In cases where the variations in the material formation times among the layers are minimal – such as sectional layers of similar cross sectional area, and high scan speed in laser-based processes, the number of layers can be used as the measure of the influence of part orientation on the manufacturing time in the optimal orientation algorithm.

8.3.2.4 Multi-criterion optimization techniques

With reference to Fig. 8.8, let different feasible part orientations be represented by points in a search space, and the axes represent the minimization criteria for the selection of the part orientation. The lowest and leftmost points in the co-ordinate system then represent the better set of solutions. In other words, orientations 1, 2 and 3 are better than the other orientations. However, it is more difficult to identify the best among orientations 1, 2 and 3.

To deal with an optimization problem with more than one criterion, three approaches are often used in practice [18]:

(1) Convert the multi-criterion optimization problem into a single objective optimization problem by summing a weighted function of various criteria.

(2) Use multiple optimization techniques, such as *min-max method, pareto analysis, goal programming*, etc., to find the optimal solution analytically.

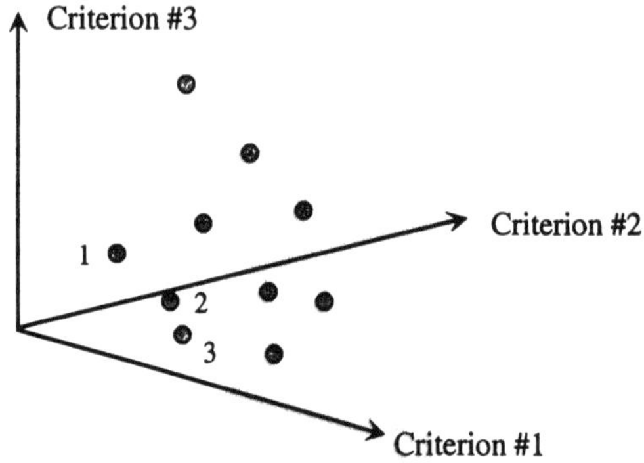

Fig. 8.8 Optimal orientation based on multiple criteria.

(3) Choose one criterion as the primary optimization objective, another criterion as the secondary optimization objective, and so on.

In the previous sections, various measures have been defined for the building inaccuracy, manufacturing time and stability of the part being built. Overhang area and the projected area of inclined planar surfaces in the part are used to evaluate the process-related inaccuracy in different orientations. The number of layers is used as the measure to evaluate the influence on the manufacturing time. The penalty function reflects the preference of an orientation from the perspective of part stability. To find a general optimal orientation, the combination of defined measures - the fabrication time, part inaccuracy and part stability - is used to form the objective function. The optimization problem is formulated as follows:

$$MIN\left\{w_s \frac{ht(\gamma)c(\gamma)}{A_{base}(\gamma)} + w_a \frac{OHA(\gamma)+INCL\_A(\gamma)}{A} + w_t \frac{NUM(\gamma)}{M\_L}\right\} \tag{8.21}$$

where $\gamma \in \Psi$, Ψ is the set of orientation candidates.

In Eq. (8.21), the denominators $A_{base}(\gamma)$, A and M_L denote the base area of the model, the total surface area of the model and the maximum number of layers when building the part in orientation γ, respectively. By introducing these denominators, the items in the objective function are made relative and dimensionless. This is to minimize their bias on the weights w_s, w_a and w_t, which indicate the relative importance of the part stability, building accuracy and fabrication time in the application. The values of w_s, w_a and w_t can be assigned by the user. When the user has equal preferences over part stability, building accuracy and fabrication time, the weights can be set to be as the same.

8.4 Direct slicing of CAD models

Various methods have been proposed for slicing of CAD models for layer-based rapid prototyping. The slicing technique translates the 3D solid model to 2D layered data. Usually, the 3D CAD model is first tessellated into an STL model. The STL model is then sent to an RP machine for slicing and data preparation. The common approach is constant thickness slicing. Where the process allows, variable-thickness or adaptive slicing techniques have also been developed[7,19,20]. By using a single type of element – the planar triangle, the STL file makes the slicing and downstream data preparation tasks easier. However, certain limitations exist while using the STL model as the interface between the CAD system and RP machine. Firstly, the STL file is inherently inaccurate as it uses planar triangles to approximate the surfaces in a CAD model. Secondly, the sequence of tessellating the CAD solid model and slicing the STL model for part fabrication is not efficient.

In contrast, if the CAD model is directly sliced in the environment of the CAD system, the aforementioned disadvantages can be eliminated or at least controlled in some way. By directly slicing the CAD model, the input data can be more accurate. The tessellation error and the data repair associated with the use of an STL model can also be avoided. Moreover, the CAD system would become more integrated with the RP facilities. An integrated and intelligent CAD/RP system, which allows the designer to have control over the design and fabrication process at the same time, would make the product development cycle more efficient and competitive. With the establishment of a common layered data format for RP, direct slicing can become more efficient. Standard or unified sliced data formats for RP have been proposed[21-23]. Design and implementation for a standard layered data format include Common Layer Interface (CLI), Layer Exchange ASCII Format (LEAF), the expansion of Hewlett Packard Graphics Language (HPGL), and certain RP machine data formats[21,23].

The following section presents an approach that directly and adaptively slices the CAD model for part fabrication without using the STL format. This is based on a (Pro/Engineer) 3D solid modeller, and a SONY SCS rapid prototyping system.

8.4.1 Adaptive thickness slicing and cusp height tolerance

The constraints of constant thickness slicing and part building in RP include:

- There exists the possibility of missing features as it is likely that when the thickness is too coarse, the sliced profiles will not retain certain fine features or the exact beginning/ending positions of some features.

- In the aforementioned case, if the thickness is made too thin to bring out the fine features, the fabrication time will be significantly lengthened. Hence, some compromise is needed between the improvement of surface roughness and the reduction of manufacturing time with constant thickness slicing.
- When dealing with a curved surface, inconsistent surface roughness will occur if a constant thickness is used to build the entire part.

For those processes that permit variable thickness control, the aforementioned constraints can be overcome with adaptive slicing. A specified cusp height tolerance can be used to maintain desired surface quality or features of the RP part. Since fine layer thickness is used only in places required by the local geometry, the number of layers can be expected to be significantly less than that with constant thickness slicing and therefore, the corresponding fabrication time. Where a desired surface cusp height gives sufficiently acceptable surface roughness, costly and time-consuming post-processing (e.g. sand blasting, CNC machining or painting/coating) to remove staircase features may also be avoided.

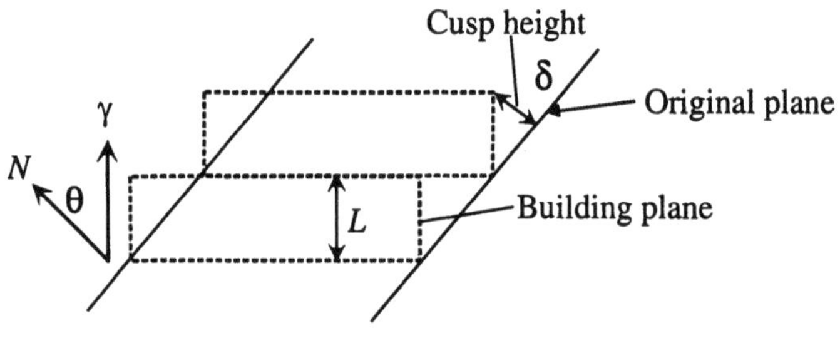

Fig. 8.9. Cusp height tolerance.

The surface cusp height is defined as the maximum distance between the actual built surface and the desired surface, as illustrated in Fig. 8.9. The surface cusp height deviation cannot be totally eliminated as long as block-like layers are used to approximate the non-perpendicular curved surface. However, by varying the layer thickness according to the surface curvature or the direction of the surface normal vector, the surface cusp height can be controlled within a user-defined tolerance.

Given a user-defined cusp height tolerance δ, if the intersecting angle between the building direction vector γ and the surface normal vector N is θ, the maximum layer thickness for building the relevant planar surface can be found as follows:

$$L = \delta / \cos\theta \tag{8.22}$$

When θ approaches 90°, the maximum allowable layer thickness can be very large. As there is a range of allowable layer thickness $[L_{min}, L_{max}]$, the actual layer thickness used to slice the referenced planar surface can be chosen within the allowable range.

8.4.2 Maximum allowable layer thickness based on curvature

For a curved surface, the surface normal vector changes from point to point. To control the layer thickness based on the local geometry, it is necessary to consider the rate of change of the tangential vector, or the surface normal curvature at that point.

8.4.3 Surface normal curvature

Assuming $\boldsymbol{r}(t)$ is an intersecting curve between a planar face and the curved surface through a point, the curvature of $\boldsymbol{r}(t)$ is defined as the normal curvature of the surface along the curve $\boldsymbol{r}(t)$. In general, the derivative of $\boldsymbol{r}(t)$

$$\boldsymbol{r}'(t) = \frac{d\boldsymbol{r}}{dt} = \frac{(\boldsymbol{r}_u du + \boldsymbol{r}_v dv)}{dt} = \boldsymbol{r}_u u' + \boldsymbol{r}_v v' \tag{8.23}$$

The curvature is:

$$k = dT / ds$$

where the unit tangential vector is:

$$\boldsymbol{T} = \frac{\boldsymbol{r}'(t)}{|\boldsymbol{r}'(t)|}$$

The equation to compute the surface normal curvature can be obtained as:

$$k = \boldsymbol{UDU}^T / \boldsymbol{UGU}^T = \boldsymbol{VDV}^T / \boldsymbol{VGV}^T \tag{8.24}$$

where $\boldsymbol{U} = (u' \;\; v');$

$$\boldsymbol{V} = (u'/v', \; 1);$$

$$\boldsymbol{G} = \begin{vmatrix} \boldsymbol{r}_u \cdot \boldsymbol{r}_u & \boldsymbol{r}_u \cdot \boldsymbol{r}_v \\ \boldsymbol{r}_u \cdot \boldsymbol{r}_v & \boldsymbol{r}_v \cdot \boldsymbol{r}_v \end{vmatrix}$$

$$\boldsymbol{D} = \begin{vmatrix} \boldsymbol{N} \cdot \boldsymbol{r}_{uu} & \boldsymbol{N} \cdot \boldsymbol{r}_{uv} \\ \boldsymbol{N} \cdot \boldsymbol{r}_{uv} & \boldsymbol{N} \cdot \boldsymbol{r}_{vv} \end{vmatrix}$$

u and v are the surface parameters, $\boldsymbol{G}$ is the first fundamental matrix of surface $\boldsymbol{r}(u,v)$, and $\boldsymbol{D}$ is the second fundamental matrix of surface $\boldsymbol{r}(u,v)$.

Because the vectors $\boldsymbol{r}_u$ and $\boldsymbol{r}_v$ are fixed at certain point on a specific surface, the ratio u'/v' uniquely determines the normal curvature in a certain direction on the tangent plane.

8.4.4 Surface normal curvature along the building direction

The surface normal curvature of interest is that along the building direction. Given a point on the surface, the unit vector γ in the building direction and the surface normal vector $\boldsymbol{N}$ at the point uniquely define a plane, denoted as the normal plane. This normal plane intersects with the surface and the intersecting curve and is denoted as $\boldsymbol{r}(t)$, as illustrated in Fig. 8.10.

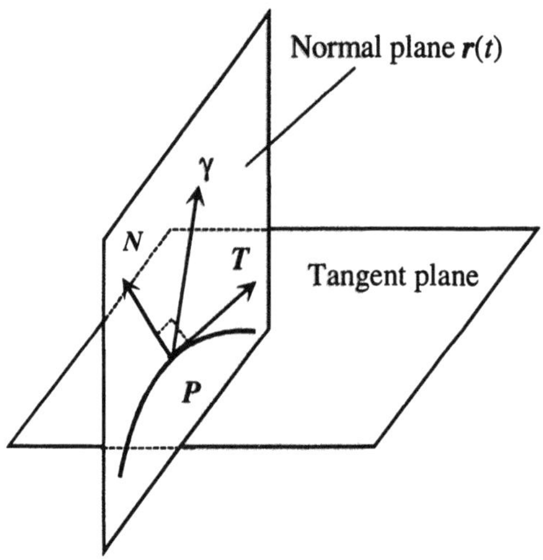

Fig. 8.10. Surface normal curvature.

The surface normal curvature along the building direction at the given point P can be obtained as follows:

(1) Find the unit tangential vector $\boldsymbol{T}$ in the direction of the intersecting curve and evaluate the unit normal vector $\boldsymbol{N}$ at point $\boldsymbol{P}$.

(2) Compute the ratio of u' to v'. By defining a vector $\overline{\boldsymbol{PH}}$ that is orthogonal to vector $\boldsymbol{T}$ on the tangent plane, as illustrated in Fig. 8.11,

$$\boldsymbol{r}'(t) = \boldsymbol{r}_u u' + \boldsymbol{r}_v v' = m\boldsymbol{T}$$

$$\Rightarrow \left(\boldsymbol{r}_u \cdot \overline{\boldsymbol{PH}}\right)u' + \left(\boldsymbol{r}_v \cdot \overline{\boldsymbol{PH}}\right)v' = m\boldsymbol{T} \cdot \overline{\boldsymbol{PH}} = 0$$

$$\Rightarrow \frac{u'}{v'} = -\frac{r_v \cdot \overline{PH}}{r_u \cdot \overline{PH}}$$

(3) Compute the normal curvature along the building direction using Eq. (8.24) based on the ratio of u'/v' and the fundamental matrix at point P.

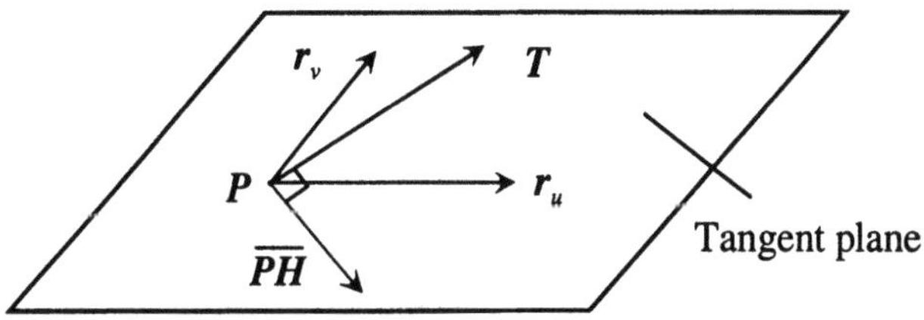

Fig. 8.11 Illustration of vectors $\boldsymbol{r_u}$, $\boldsymbol{r_v}$ and unit vector $\boldsymbol{T}$.

8.4.5 Computing the maximum layer thickness at a point on the surface

Based on the surface normal curvature in the building direction and the surface normal vector at point P on the surface, the maximal allowable layer thickness can be found for a given cusp height tolerance. Considering layers built from the bottom up, Eq. (8.25) can be derived [24] for different cases shown in Fig. 8.12:

$$\text{(a) } L = -\rho\cos\theta + \sqrt{\rho^2\cos^2\theta + 2\rho\delta + \delta^2}$$
$$\text{(b) } L = \rho\cos\theta + \sqrt{\rho^2\cos^2\theta + 2\rho\delta + \delta^2}$$
$$\text{(c) } L = -\rho\cos\theta - \sqrt{\rho^2\cos^2\theta - 2\rho\delta + \delta^2}$$
$$\text{(d) } L = \rho\cos\theta - \sqrt{\rho^2\cos^2\theta - 2\rho\delta + \delta^2} \qquad (8.25)$$

where ρ is the radius of the curvature circle at point P.

8.4.6 Optimal layer thickness at reference height

The preceding section shows computational methods to determine the maximum allowable layer thickness at a given point on a curved surface. At a given height, there would be varying contours bounding the section in that slice. It is necessary to find the optimal layer thickness at the referenced height level by searching among points along the slice contours.

8.4.6.1 One-dimension search problem

Given the parametric description of a surface, the problem to find the optimal layer thickness for the surface at height Z can be formulated as a constrained optimization problem:

$$Min\{L(\boldsymbol{r}(u,v),\delta),\ \text{for all}\ \boldsymbol{r}(u,v)\cdot\gamma = Z\} \tag{8.26}$$

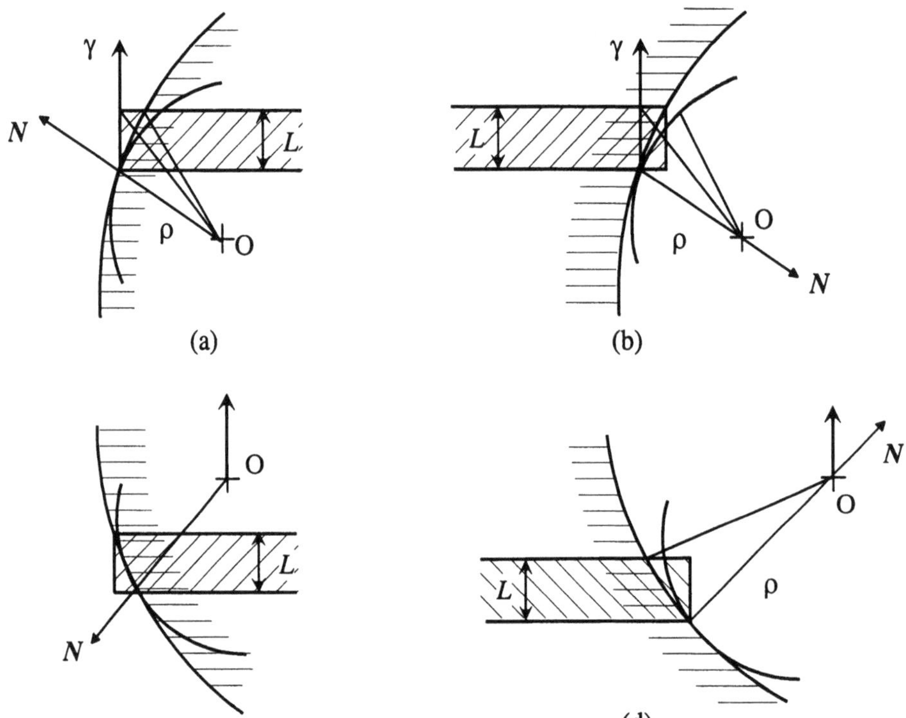

Fig. 8.12 Maximum layer thickness for different geometric configurations.

$L(\boldsymbol{r}(u,v), \delta)$ represents the computed layer thickness at point $\boldsymbol{r}(u,v)$ with the given cusp height tolerance δ. Analytical methods can be used to find the solution for this optimization problem when there exists a single explicit surface equation $\boldsymbol{r}(u,v)$ and the equation is not complex. Otherwise, mathematical programming techniques can be employed to search for the optimum. But when the surface involved is non-continuous, or multiple surfaces are involved in the cross-section, both analytical and mathematical programming methods may fail to find a feasible solution. A random search method has been found to be able to find the optimum in such cases.

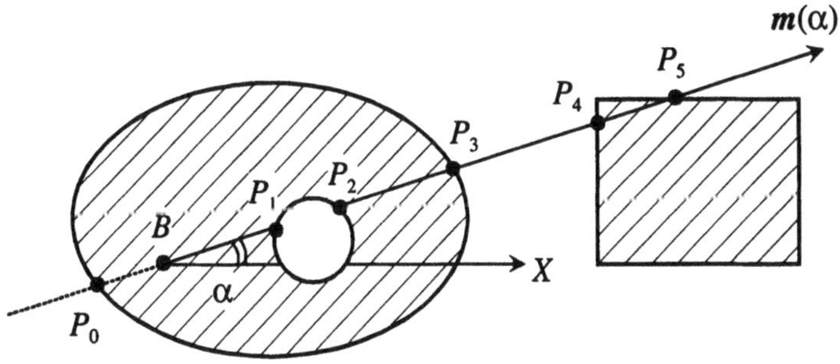

Fig. 8.13 Cross section at referenced height Z.

Consider the cross section at height Z. Assume that B is a point in the interior of the cross section, as illustrated in Fig. 8.13. A series of lines can be drawn from point B. Each line may have several intersecting points with the boundaries of the part. If point B is used as the origin to construct a 2D polar coordinate system in the plane, the line vector can be described using angle α. By changing the angle α to choose different directional line vectors, points on the part boundaries are sampled for the computation of the allowable layer thickness. Eq. (8.26) is then modified into a one-dimensional search problem:

$$Min\ \{L(\boldsymbol{P}_i(\alpha), \delta)\ ,\ i = 0, 1, 2\ ,\ldots\ldots\ ,\ n\text{-}1\} \tag{8.27}$$

where $0 \le \alpha \le 180$, n is the number of intersecting points between line vector $\boldsymbol{m}(\alpha)$ and the boundaries of part, and P_i (i = 0, 1,......, n-1) are the intersecting points.

8.4.6.2 Genetic algorithm for one-dimensional search

Proposed in the early 70's, genetic algorithm (GA) has been known for its performance and robustness in the optimization of 'bad functions'. 'Bad functions' refer to those functions that have a discontinuous, multi-modal or noisy domain. Unlike traditional search methods, such as Fibonacci search and steepest descent search which use derivatives to find a local extremum, genetic algorithm randomly updates the set of selected points and discards unfavorable solutions gradually in the search. In general, GA is more efficient on finding a global optimum than the traditional optimal search methods[25].

Basically, GA mimicks the natural (gene) selection phenomenon to find the optimal individual by improving the fitness of individuals through generations. Some key terms related to GA include: *Population* refers to all possible individuals. In GA, each individual is called a genotype. Usually the population has a fixed size as it evolves from generation to generation. All possible different genotypes are collectively called the *genetic pool. Fitness* defines the measure of performance with each genotype. High fitness means greater chance for survival with the related genotype through the evolution. The seed with high fitness thus tends towards the optimal value in the selection process. *Genetic operators* are defined manipulations for the reproduction of offspring. The main genetic operators include *crossover, inversion* and *mutation.*

In GA, the major task is to find a suitable representation of solutions. In the case of the aforementioned one-dimensional optimization problem, this task is relatively simple. The angle parameter α can be readily coded into a binary number if the requirement on the resolution is given. Assume the resolution is 0.5° over a range of 0~180°, then a 9-digit binary number is sufficient to represent all possible solutions.

The fitness function is defined as:

$$\text{fitness} = MAX\_VALUE - L(\alpha) \tag{8.28}$$

where $L(\alpha)$ is the minimum value of all those computed layer thickness at different points $P_i(\alpha)$, which are the intersecting points between the line vector $\boldsymbol{m}(\alpha)$ and the boundary surfaces. *MAX_VALUE* is a given large number, for the purpose of making the candidate with larger fitness to represent the better solution.

Three operations, namely, *crossover, mutation* and *inversion*, are used in the reproduction process. In order to find the optimum value, in each generation, the best seed will be noted and preserved during the follow-up reproduction process. The best seed is dynamically changed when a member with larger fitness value appears in the new generations. The search process is terminated when the best seed in the population has not changed after five consecutive generations. The final best value is then chosen as the optimum. The corresponding layer thickness can be computed from the fitness definition based on this value.

Having determined the minimum layer thickness appropriate for each layer through GA search, the computed values are compared with the layer thickness constraints for the specific RP machine. These are the allowable minimum and maximum thickness $[L_{min}, L_{max}]$, and the minimum possible incremental change in thickness ΔL.

8.4.7 Implementation of adaptive slicing algorithm

The aforementioned adaptive algorithm has been implemented on a solid modeling environment (Pro/Engineer) for parts built using a specially modified LL machine (Solid Creation System from SONY)[12]. The building layer thickness of the LL machine cannot be larger than 0.3mm or less than 0.05mm. The layer thickness must be in increments of 0.01mm. Hence, the computed layer thickness must be adjusted within the range of [0.05mm, 0.3mm] in increments of 0.01mm. Fig. 8.14 shows the flowchart of the implemented adaptive slicing algorithm. The segmentation and the slicing of the computer-aided design (CAD) solid model are illustrated in Fig. 8.15.

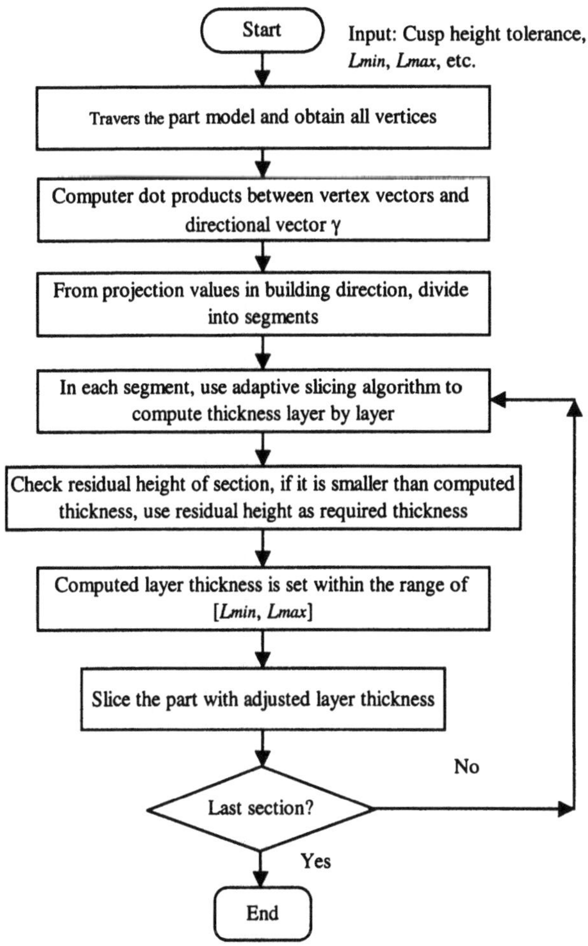

Fig. 8.14 Flowchart of the implemented adaptive thickness slicer.

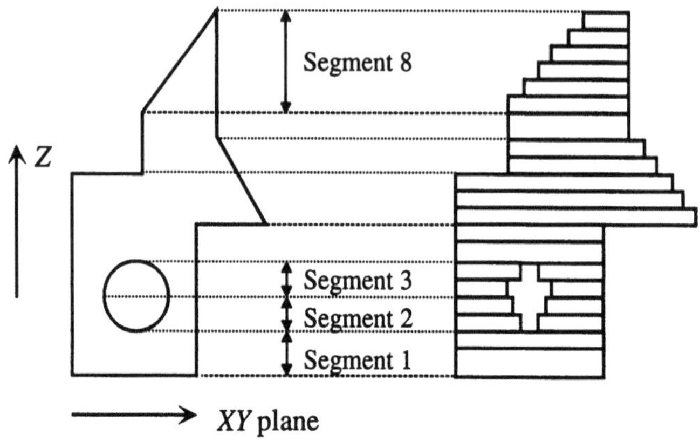

Fig. 8.15 Segmentation and slicing of the CAD model.

To show the effect of the adaptive slicing algorithm, a wine glass shown in Fig. 8.16 has been designed and sliced in the CAD environment and the sliced data subsequently output to the SCS machine. A plot of the change in layer thickness against the height in number of layers is shown in Fig. 8.17. It can be seen that the part is essentially divided into four segments. The first segment from the bottom of the glass covers the first layer to the 115th layer. The second segment is the part of cylindrical pole. Large layer thickness – up to the allowable maximum of 0.3mm - is used in this second segment. The third segment involves the fillet surface between the cylindrical pole and the base of the revolved surface. This third segment includes the layers between the 220th layer and the 300th layer. Thinner layer thickness has been used to slice the third segment due to the relatively large normal curvature of the fillet surface. The last segment of the glass is formed by two revolved surfaces. From Fig. 8.17, it can be seen that the surfaces gradually become perpendicular to the building plane as Z increases. The computed layer thickness in this segment thus increases to the maximum with the layer number. During the process of computing the layer thickness, the last two layers in each segment may need to be specially adjusted. If the remaining height of a segment is less than the allowable minimum of 0.05mm, the previously computed layer thickness will have to be adjusted to incorporate the remaining height to ensure the required height in that segment is attained.

This example shows that the slicing algorithm identifies the local geometry requirements and automatically applies a smaller layer thickness to build the more curved sections in the part. Fig. 8.18 compares of the number of layers to slice the glass by constant thickness slicing and adaptive slicing algorithms for the surface cusp height tolerances set at 0.05mm and 0.1mm, respectively. Using adaptive layer thickness in the building process reduces the part fabrication time significantly due to the reduced number of layers required, while the surface cusp height is kept within the tolerance.

Fig. 8.16 Shaded view of glass model for testing adaptive slicing.

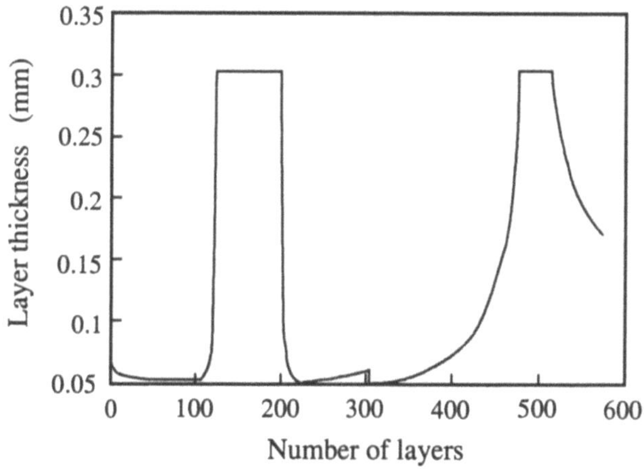

Fig. 8.17 Computed layer thickness of the glass.

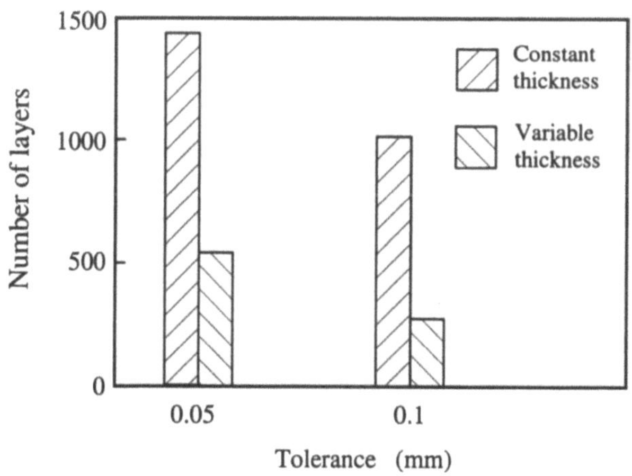

Fig. 8.18 Comparison of number of layers of the glass.

8.5 Decision support for process optimization and selection

As can be seen from the preceding sections, RP selection criteria, such as dimensional accuracy, surface roughness, manufacturing time and cost, are complex functions of a number of factors in the fabrication process. These factors include the geometric characteristics of the model, the building orientation, the material used in the process, the building mechanism, the building style and other process parameters. Thus, the selection of RP material/machine requires comprehensive knowledge and expertise on low-level material attributes and machine performance.

To facilitate the decision-making process, different types of information must be gathered and organized efficiently. In general, the information involved can be categorized in the following five fields:

(1) RP material data, including mechanical, thermal and chemical properties of the material, its unit cost, appearance, etc.;

(2) RP machine data, including material types, building styles, controllable process parameters, machine capabilities, etc.;

(3) RP application data, providing typical application-related requirements on material property, tolerance, surface roughness, etc.;

(4) Geometric data, providing information such as the part envelope, part volume, smallest linear dimensions, types of form features, etc.;

(5) Model specifications, including application purpose, dimensional tolerance, general surface roughness, material properties, etc.

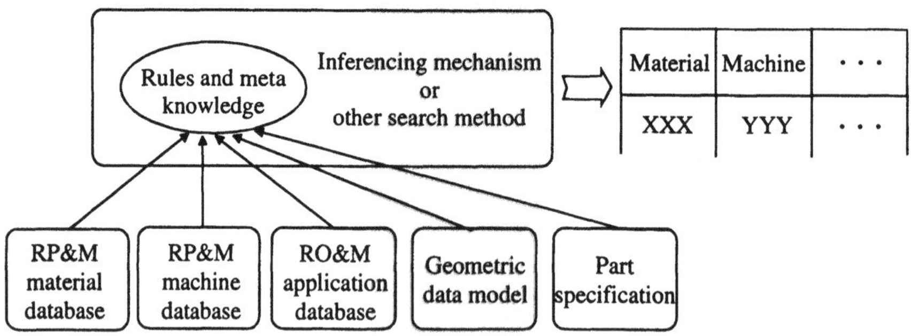

Fig. 8.19 Information requirements in RP process selection.

These five fields define the five highest-level clusters of information needed in RP material/machine selection. The evaluation of building quality, the computation of manufacturing time and cost require access to these five clusters of information. Based on the organized information and proper models, appropriate decisions can be made for the fabrication of RP parts, as illustrated in Fig. 8.19.

8.5.1 Knowledge-based RP material/machine selection

It is obvious that the selection of the appropriate RP material/machine is a knowledge-intensive process, requiring understanding and knowledge of the different RP processes. When RP experts select a process for a product design, the design requirements or the part specifications are usually abstracted and refined into low-level goals first. Since the objectives required of an RP part may be conflicting when detailed into low-level tasks, meta-knowledge or a set of rules is often needed to guide the decision flow in the selection. The interdependent relationships among the manufacturing quality, the manufacturing time/cost and the part geometry, as well as the unstructured characteristic of the process, suggest the suitability of employing a knowledge-based approach in the selection of an appropriate RP material/machine.

8.5.2 Knowledge-based systems and RP part fabrication

A knowledge-based system (KBS) is defined as a software system with structured knowledge about a field of expertise and ability to solve problems within a specific domain by using knowledge derived from experts in the field [26]. As illustrated in Fig. 8.20, a knowledge-based system usually consists of databases, domain knowledge, control/monitor mechanisms, inferencing, computing & reasoning modules, user interface, database interface, and reporting system. The domain knowledge can be represented in various forms, e.g., symbols, production rules, algorithms, decision trees, theorems, etc.

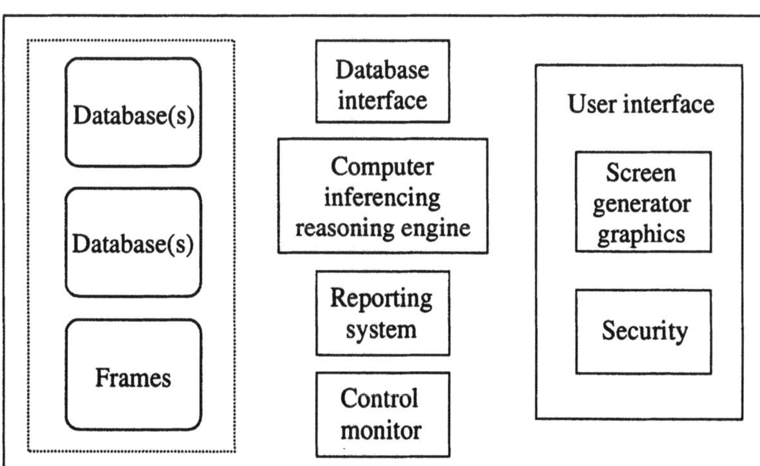

Fig. 8.20 Anatomy of a knowledge-based system.

For design and manufacturing applications, various intelligent knowledge-based systems have been developed in recent decades to address the problems of integration, data management, process planning, control and communication at varying levels [27,28]. These projects aimed to increase automation and include intelligence in conventional manufacturing processes.

An object-oriented knowledge-based RP material/machine selector has been developed [12] that given a CAD model, identifies suitable RP materials and machines in response to the part geometry and application requirements on dimensional accuracy, surface roughness and mechanical or thermal properties. Object-oriented databases are used to organize the process-relevant information, such as machines, material and application cases. Based on the architecture of the knowledge-based RP material/machine selector and developed modules on optimal orientation and direct slicing, an integrated decision support system has also been implemented for part fabrication with RP.

8.6 Summary

The introduction of RP technology has expanded the scope and capabilities of modern manufacturing technologies. Unlike conventional material removal manufacturing processes, the material additive principle of RP gives the manufacturers opportunities to manipulate the micro-characteristics and internal structures of fabricated parts. In this chapter, various process-planning issues relevant to the RP part fabrication processes have been identified and discussed with respect to four relatively common processes – LL, SLS, LOM, and FDM. Understanding of these pertinent issues is essential in order to make informed selection of an appropriate RP process, machine and material for specific prototyping requirements. Appropriate knowledge-based decision support tools have also been briefly presented that aim to facilitate the RP process, machine and material selection.

8.7 References

1. D.J. Whitehouse, *Handbook of Surface Metrology*. Institute of Physics Publishing, Bristol, UK, 1994.
2. P.E. Reeves, and R.C. Cobb, *Proc. of the 5th Eur. Confer. on RP.*, Helsinki, Finland. July 4th-6th 1996, 59.
3. I.Y. Tumer, D.C. Thompson, K.L. Wood and R.H. Crawford, *J. Manuf. Syst.*, Vol. 17, (1) (1998), 23.
4. S. Allen, and D. Dutta, J. Design and Manuf., No. 5 (1996), 153.
5. G. Boothroyd, *Fundamentals of Metal Machining and Machine Tools*, Washington, Scripta Book Corp., 1975.

6. S. Allen, and D. Dutta, *"On the Computation of Part Orientation Using Support Structures in Layered Manufacturing"*, Technical Report, 1994, Dept. of Mechanical Engineering, University of Michigan.
7. N.S. Puduhai, and D. Dutta, *"Determination of optimal orientation based on variable slicing thickness in layered manufacturing"*, Technical Report UM-MEAN-94-14, 1994, Department of Mechanical Engineering, University of Michigan, Ann Arbor, MI.
8. J. Hur, and K. Lee, *"Determination of Optimal Part Orientation for Stereolithographic Rapid Prototyping"*, Technical Report, 1995, Department of Mechanical Design & Production Engineering, Seoul National University.
9. W. Cheng, J.Y.H. Fuh, A.Y.C. Nee, Y.S. Wong, H.T. Loh, and T. Miyazawa, *RP. J.*, Vol. 1, (4) (1995), 12.
10. P.T. Lan, S.Y. Chou, L.L. Chen, and D. Gemmill, *Computer-Aided Design*, Vol. 29, (1) 1997, 53.
11. D. Frank, and G.J. Fadel, *J. Intelligent Manuf.*, No.6 (1995), 339.
12. F. Xu, *"Integrated Decision Support for Part Fabrication with Rapid Prototyping and Manufacturing Systems"* Ph.D. Thesis, Mechanical Engineering Department, National University of Singapore, 2000.
13. T. Woo, and R. Liang, *Computer Aided Design*, Vol. 25, (4) (1993), 233.
14. K.F. Roth, Mathematika Vol. 1 (1954), 73.
15. J.M. Hammersley, *Annals of the New York Acad. of Sci.*, Vol. 86 (1960), 844.
16. P. Preparata, Franco and Shamos, Michael Ian, *Computational Geometry: an Introduction*, Springer-Verlag, New York, 1985.
17. K. Mulmuley, Computational Geometry: *an Introduction through Randomized Algorithms*, Prentice-Hall, 1994.
18. Andrzej Osyczka, *Multicriterion Optimizationn in Engineering*, Ellis Horwood Ltd. 1984.
19. A. Dolenc, and I. Makela, *Computer-Aided Design*, Vol. 26, (2) (1994), 119.
20. P. Kulkarni, and D. Dutta, *Computer-Aided Design*, Vol. 28, (9) (1996), 683.
21. A. Dolenc, and I. Makela, *Proc. of the 1st Eur. Confer. on RP.*, 1992, 155.
22. R. Jamieson, and H. Hacker, *PR. J.*, Vol.1, (2) 1995, 4.
23. V. Kumar, and D. Dutta, *"An assessment of data formats for layered manufacturing"*, Report No. UM-MEAM-95-22, University of Michigan, Ann Arbor, 1995.
24. F. Xu, Y.S. Wong, H.T. Loh, J.Y.H. Fuh and T. Miyazawa, *RP. J.*, Vol. 3, (3) (1997), 76.
25. D. Ginat, "Genetic algorithm - a function optimizer", *Technical Report*, Dept. of Computer Science, U. of Maryland, 1988.
26. M.A. Carrico, E. Girard John and J.P. Jeniffer, *Building Knowledge Systems*, Intertext Publications/Multiscience Press, Inc., 1989.
27. A. Kusiak, eds., *Knowledge Based Systems in Manufacturing*, Taylor & Francis, Philadelphia, PA, 1989.
28. Y.M. Chen, R.A. Miller, and Sevenler, Korhan, *J. Intelligent Manuf.*, No. 6 (1995), 321.

Chapter 9

RAPID TOOLING AND ITS APPLICATIONS

9.1 Rapid tooling development

Rapid tooling (RT) or rapid manufacturing (RM, i.e. rapid manufacturing due to fast fabrication of molds or tools or direct fabrication of functional prototypes) can be accomplished through the use of rapid prototyping processes followed by some subsequent processes [1]. An RP model can be used as a pattern to investment cast the actual part; and an injection mold (mainly cavity and core inserts) can be designed and fabricated based on the pattern to manufacture the final parts. RT is a natural technological extension of RP. Current RP technologies are neither able to prototype products in a wide range of materials, nor well suited to manufacture a large batch of parts. This has led to the adoption of various tooling options; such options and techniques are collectively termed RT [2].

Today, emphasis in production tooling has shifted to *Rapid Tooling (RT)* due to the advance of RP technology. However, applying RP to RT has one major inadequacy - most RP systems are not able to fabricate functional prototypes. The materials that can be used in the RP processes to fabricate the prototype presently do not possess the same physical, mechanical, thermal or electrical properties of the end material to be used in the manufacture of the final part. Clearly, there needs to be a way to rapidly manufacture molds that can make prototypes with the same end material. Secondly, and perhaps the more importantly, RT promises to greatly reduce the lead time to market a product. Production tooling has been traditionally fabricated by highly skilled tool and mold makers using CNC machining and EDM. Such conventional tooling processes are very mature and there is a wide selection of suppliers with a good range of hardened tools and standard mold bases [3]. However, it takes at least 4 to 8 weeks to manufacture a conventional mold. In contrast, the same mold can be manufactured in less than 2 weeks using RT techniques. With reduced lead time, rapid tools can batch manufacture enough parts for an initial market launch, seizing market share while conventional molds for the final mass manufacturing are concurrently engineered.

9.2 Rapid tooling techniques and applications

The fabricated tools usually have varying mechanical (e.g. hardness, wear resistance, roughness, etc.), thermal, electrical and chemical properties that are dependent on the materials and the processes used in the RT technique. Hence, it is essential to have a good understanding of the RT techniques identified for specific tooling applications. Although different RT techniques (e.g. soft tooling vs. hard tooling, indirect tooling vs. direct tooling, or prototype tooling vs. production tooling) can be used to fabricate various tools for injection molding, investment casting, die casting, sand casting, etc, the following sections focus only on RT techniques suitable for plastic injection molding.

Generally, RT techniques can be categorized as either *direct* or *indirect* tooling. Direct tooling requires no intermediate step to fabricate the tools. The tools are formed directly by the RP process. Indirect tooling begins with a RP-fabricated master prototype. Tools are then formed according to the shape of the prototype, either by casting or by coating processes. Fig. 9.1 shows the classification and main types of various RT techniques.

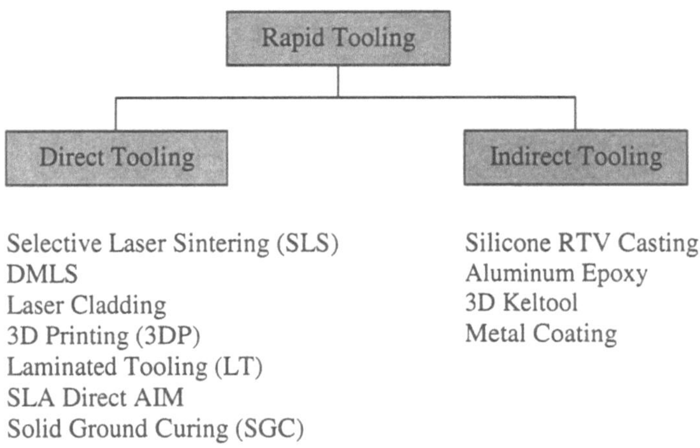

Fig. 9.1 Main types of RT techniques.

Rapid Tooling applications include:

- Rapid plastic injection mold/die;
- Rapid functional parts; and
- Rapid EDM electrodes.

The advantage of manufacturing rapid mold using RP techniques is that conformal cooling channels, which are designed with complex geometries not amenable to be shaped by machining, can be incorporated in the mold inserts. The details will be presented by a case study in the later section of this chapter. RT can significantly reduce the tooling development cycle time and cost. Savings of 50% in time and cost are often reported by companies employing RT [2]. With the ability to use various materials in RT, new and advanced-functional parts are likely to be built. EDM electrodes with complex geometry can be also made much faster by using various RT technologies to reduce the time and cost of fabricating tooling.

9.2.1 Direct tooling

Direct tooling pose a great challenge to RP/RT applications and will become a major R&D focus in the next few years. The following are a few laser-based direct tooling methods:

RapidTool by SLS

DTM's RapidTool® [4] process uses RapidSteel® 2.0 stainless steel material to form 3D tooling parts by SLS process (see Section 5.4 and 5.5 for details on its process and material used). Essentially, RapidSteel® 2.0 powder is steel powder coated with polymer binder. The "green' part is then removed and placed in a furnace. An initial furnace cycle debinds the steel powder. The resulting "brown" part is then set on an alumina plate, surrounded by bronze ingots within a graphite crucible, and placed in a furnace. The heated bronze melts and infiltrates the part by capillary action. Compared to RapidSteel 1.0 (carbon steel powder) the infiltrant material is bronze instead of copper. The binder system was also modified. The dimension inaccuracy due to shrinkage can be as small as 0.2% as reported.

Although the part is strengthened by infiltration, there may be unexpected amount of shrinkage during the furnace operation. In the Copper Polyamide™ (PA, a mixture of copper and nylon powder) process, only the shell with a certain thickness is generated instead of sintering the whole solid block. No furnace operation is necessary and the inserts are machinable. Hence, the inserts have to be back filled by aluminum epoxy resins. Then it can be fitted into a mold base for injection molding. Copper Polyamide technology will allow short-run production (several hundred parts) in carbonate, polypropylene, and ABS for functional testing and concept evaluation. Fig. 9.2 shows an injection tooling built by SLS process.

Fig. 9.2 Copper PA tools created by SLS process (Courtesy of DTM Corporation [4]).

DirectMetal and DirectSteel by DMLS

EOS's DirectTool™ process uses its Direct Metal Laser Sintering (DMLS) technique [5] to sinter metal powder. Its powder system is the mixture of nickel, copper-phosphate and bronze. The process mechanism is similar to the SLS process. However, unlike the SLS process, the process requires no pre-heating of the metal powder. Sintering is achieved by the use of a 200W-CO_2 laser system. Presently, EOS offers two types of powder for the EOSINT M machines. The first generation powder (DirectMetal) is bronze-based while the second generation powder (DirectSteel) is steel-based. The more detailed descriptions on the EOS process and material can be found in Sec. 5.4.3.

For the bronze-based powder, sintering takes place in normal atmosphere; no inert gas fills the building chamber whilst for the second material, an integrated nitrogen generator provides an inert atmosphere for the sintering process. Parts made with the bronze-based powder are porous and an infiltration process with high temperature epoxy resin is necessary. Using epoxy resin eliminates the need for furnacing which will otherwise affect the accuracy of the parts. Injection molding tools using epoxy-infiltrated core and cavity inserts can produce up to thousands of parts for some standard plastic materials while using proper injection molding parameters. The details on the process will be described in Sec. 9.3. Fig. 9.3 shows the core and cavity inserts created by the DirectTool process.

Fig. 9.3 Plastic injection tools, cavity and core built by DMLS process (Courtesy of EOS GmbH [5]).

Laser Cladding

Processes such as Laser Engineered Net Shaping (LENS™) [6], Shape Deposition Modeling (SDM) [7], Laser Generating (LG) [8], Controlled Metal Build-up (CMB) [9], etc. are variants of the laser cladding technique. Instead of having a bed of metal powder, laser cladding processes deliver metal powder via a nozzle. A high power laser such as a Nd:YAG laser is used to melt an area on a metal substrate while the nozzle simultaneously delivers metal powder to the molten weld pool. To prevent oxidation, the build chamber is filled with an inert gas. The process details of laser cladding can be found in Sec. 6.1.1.

Processes like SDM employ a milling operation to ensure a flat upper surface before depositing the next layer. The advantage of laser cladding is that by melting the metal powder, a fully-densed metal part and tooling can be formed directly without post-processing.

Laminated Tooling (LT)

In LT, the tools are built by stacking metal laminates together. First, the thickness of the laminate to be used is determined. Suitable software slices each half of the mould at the pre-determined thickness either horizontally or vertically, depending on the part. Laser cutting or water-jet cutting is employed to cut the required contour of each laminate. Burrs are removed and the laminates are thoroughly cleaned before they are stacked and held together (by spot welding, nuts and bolts, adhesives etc.) in the order to form the mold. Finishing operation can be done by CNC or EDM machining. The application of the LT [10] is currently limited by the poor surface finish of the assembled tool. This is caused by staircase formed by the laminated edge.

9.2.2 Indirect tooling

Silicone rubber molding

Also known as *soft tooling*, silicone casting is a very popular process for the fabrication of rapid prototypes. Due to high elasticity of silicone, parts with simple undercuts can be cast and removed easily. Silicone molds are able to inject up to 20 parts of Room-temperature Injection Material (RIM), such as Polyurethane (PU). The process begins with a well-polished master prototype. Liquid silicone and a hardener is mixed and poured into a mold containing the master prototype. After curing in an oven, the master prototype is removed by cutting along the parting line (Fig. 9.4). Fig. 9.5 shows a set of silicone-made cavity and core mold to produce products in different materials.

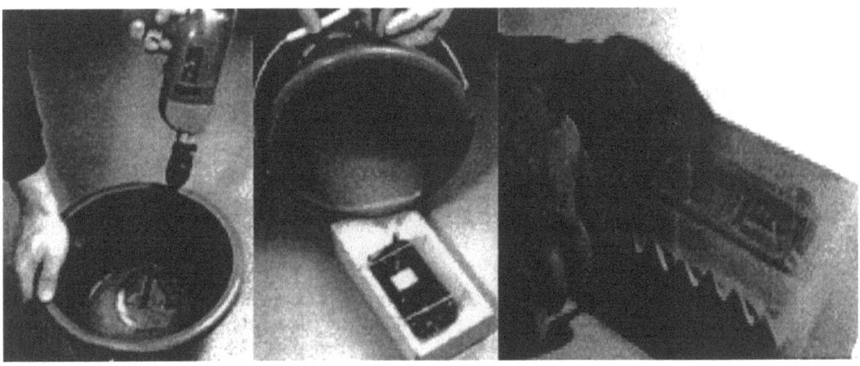

Fig. 9.4 Mixing, casting and demolding of silicone mold.

Fig. 9.5 Silicone molds for vacuum casting [2].

Aluminum Epoxy Casting

The material used for this process is a mixture of aluminum powder and epoxy resin (i.e. 35% epoxy and 65% aluminum composite). Similar to silicone mold casting, the process also begins with a well-polished master prototype. Aluminum-epoxy resin is mixed with a hardener and poured into a mold containing the master prototype, forming one half of the mold. The resin is allowed to cure. With the master prototype still in the first half, the part is inverted and parting boards are formed around it. Resin is poured over the other half to complete the casting process. The part is allowed to cure in an oven before de-molding. The tool life is longer compared to soft tooling such as silicone mold.

3D Keltool

This proprietary process by 3D Systems also begins with a master prototype. The prototype is used to create an intermediate silicone mold. Proprietary metal particles and binder material are mixed and then poured into the silicone mold. After curing, the "green" metal part is demolded and fired at high temperature to burn off the binder and fuse the metal particles. Subsequently, it is infiltrated with copper. Fig. 9.6 show the mold inserts built by 3D Keltool process [11].to mass produce plastic parts. The metal inserts have approximate 70% steel and 30% copper and can withstand a higher molding pressure and temperature (greater than 650 degree C). The tool has therefore been used to die cast zinc and aluminum.

Fig. 9.6 3D Keltool mold inserts [11] (Courtesy of 3D Systems).

Metal Coating

Metal coating can be achieved by several methods, such as electroplating, arc-spray, plasma spray, etc. Typically, the process begins with a master prototype. The master may be produced by LL, LOM or other suitable RP processes. Next, a layer of metal (usually nickel, copper or zinc) is deposited on the surface of the master prototype. The layer is subsequently strengthened by a backing material such as aluminum-epoxy. The process is repeated for the other side of the master prototype to obtain both cavity and core parts.

9.2.3 Comparison of RT techniques

Selecting an appropriate RT process is not an easy task. If the aim is to inject a large number of plastic parts, a mold with high strength and wear resistance is necessary. In this instance, metal molds are the most appropriate. They are also desirable because of better thermal conductivity. However, the accuracy of metal molds is difficult to control. The sintering of metal powder depends on a combination of factors such as the oxidation, humidity, laser optical calibration, powder size and compactness, etc. In addition, processes such as SLS, 3DP and 3D Keltool would require furnacing cycles to debind the polymer binders and replace them with another metal. Such furnacing cycles are often difficult to control and will inevitably further affect the accuracy of the mold.

When the number of parts needed are less than 100, it would be more economical to consider photopolymer resin molds. The accuracy of such molds is easier to control. On average, photopolymer resin molds can inject up to 100 parts before they start to wear. To prolong the life of the photopolymer molds, they can be treated with a coating of metal that improves the surface hardness and finish. Generally, injection molds made of photopolymer resins must be run at relatively low temperature because the resin starts to deform at about 70°-90° C. Moreover, because such resins lose heat very slowly, the molds must be run with very long cycle time, typically seven to ten minutes [12].

Processes such as silicone rubber molding and aluminum epoxy casting are relatively fast and cheap ways to manufacture prototypes. Silicone molding is a soft tooling method and is generally not used for injection molding. However, research has shown that metal-filled silicone has been able to inject up to 30 parts with polypropylene and polyamide [13]. Aluminum-epoxy resin having glass transition temperatures below that of common engineering plastics would require longer cycle time between shots to cool down.

Laminated tooling and metal coating techniques can be achieved if appropriate fabrication machines are available. However, such processes usually require the necessary experience before getting the process parameters right. A relative comparison of various major RT techniques is given in Table 9.1 below.

Table 9.1 Relative comparison of various RT techniques

Direct Tooling						
RT Technique	Material	No. of parts injected	Accuracy (mm)	Hardness	Max part size (mm^3)	Layer thickness (mm)
SLS	Steel/-copper	100,000	0.25	Good 22HRC	381x330 x457	0.075
EOS	Bronze based & Steel based	10,000	0.1	Fair 60HRV 60HRB	250x250 x150	0.05
LENS	Metal	n.a.	0.13	n.a.	300x300 x300	0.127
3DP	Metal	n.a.	n.a.	n.a.	203x250 x203	n.a
SDM	Metal	n.a.	n.a.	n.a	n.a.	n.a.
Laminated Tooling	Metal	n.a.	n.a.	n.a	No upper limit	n.a.
SLA	Photo-polymer	100	0.125	Poor	500x500 x585	0.1
SGC	Photo-polymer	100	0.1	Poor	500x350 x500	0.1
Indirect Tooling						
RT Technique	Material	No. of parts injected	Accuracy (mm)	Hardness	Max part size (mm^3)	Layer thickness (mm)
Silicone	Silicone	20 (RIM)	n.a.	n.a.	No upper limit	-
Al-Epoxy	Al/epoxy	500	n.a.	n.a.	No upper limit	-
3D Keltool	Steel/ copper	1 million	0.125	35-55 HRC	25x25 x25	-
Metal Coating	Metal coating	1,000	n.a.	n.a.	n.a.	-

9.3 RT for injection molding – a case study [14, 15]

Injection molding is the most common process to manufacture plastic parts in materials such as ABS, polycarbonate, nylon, etc. The process involves the use of tooling called injection molds (see Fig. 9.7) that comprise of the cavity and core inserts, mold base components, ejector pins, cooling channels, sliders and lifters, etc. The main components are the cavity and core inserts (see Fig. 9.8), the tooling components that form the shapes of the injected parts (or molding).

Manufacturing the complex cavity and core quickly and economically is crucial factor to reduce the lead-time of mold fabrication. When the cavity and core possess free-form features, conventional methods involve many machining processes, such as CNC, grinding, EDM and wire-cut. EDM and wire-cut processes are expensive and time consuming. With the appearance of RP technology, it becomes easier to realize free-form fabrication. On the other hand, the arrangement of cooling channel in injection molds directly influences the quality of the products and injection cycle time. Conformal cooling channel, due to its more even cooling effect, is the most ideal option. It can greatly shorten the injection cycle time and reduce the residual stresses of the final parts. Conventional fabrication methods impose severe limitations on the configuration of the cooling channels. The layered manufacturing feature of the RP technology allows for the fabrication of molds with complex conformal cooling channel [17]. The following case study presents a complete process using RP/RT technique in building an injection mold. The performance of the injection mold is also evaluated and compared.

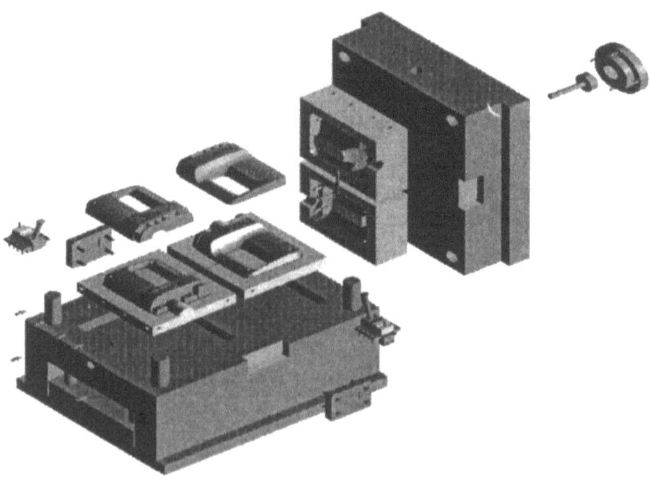

Fig. 9.7 An injection mould (Courtesy of IMOLD™ [16], ManuSoft Plastic Pte. Ltd.).

Fig. 9.8 The cavity and core for a phone cover (Courtesy of IMOLD^{TM}).

9.3.1 Design and fabrication processes

The process of mold design and fabrication based on RP/RT is shown in Fig. 9.9. The detailed steps are listed below:

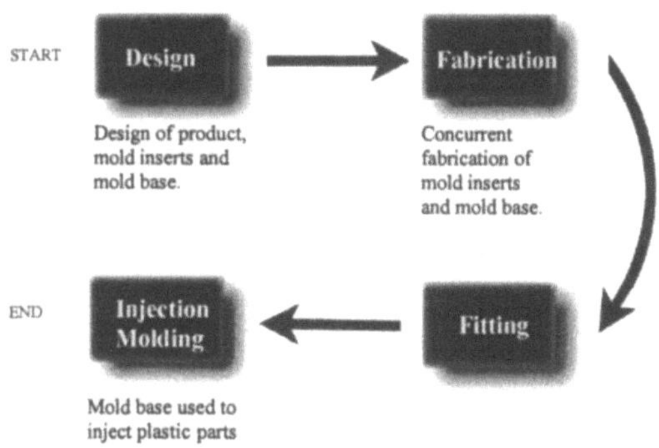

Fig. 9.9 Process of creating mold using RP/RT.

Step 1: Design

The process begins with the design of a product (Fig. 9.10) on a CAD system. In order to visualize and verify the design, a RP part needs to be built. After that, CAD models for the mold cavity and core inserts are generated, followed by design of the mold base assembly. The wall thickness of the case study part (a cup) is 2 mm; it is considerably thick compared with ordinary plastic products. Therefore, the arrangement of cooling channel must be carefully considered. Otherwise, the molding quality is difficult to control. On the other hand, due to the depth of the cup (75mm), efficient cooling management is highly desirable. Conformal cooling channels in the form of spirals have therefore been incorporated into the mold inserts design (as shown in Fig. 9.8).

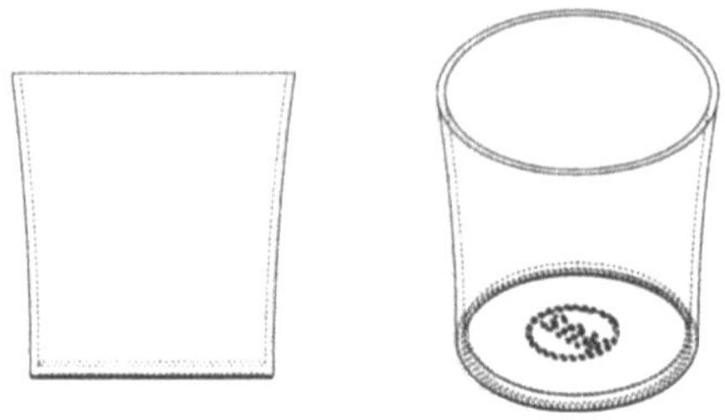

Fig. 9.10 Cup design.

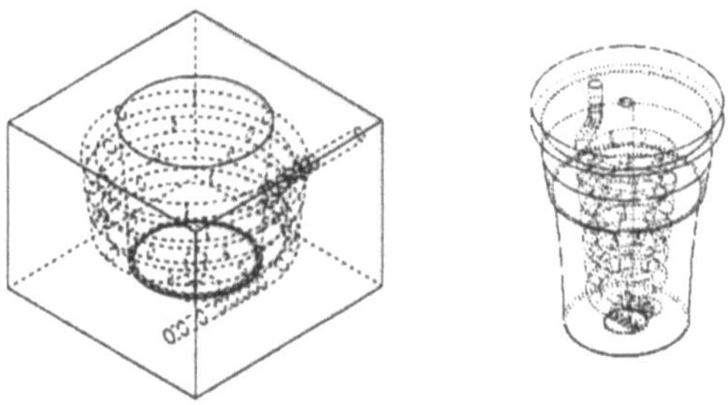

Fig. 9.11 Core and cavity inserts with conformal cooling.

To ensure accurate fitting between the cavity and mold cavity plate, the cavity insert is expanded by 2 mm in the *x*, *y*, and *z* directions. This will ensure sufficient allowance to trim the sides to the desired dimension after inevitable shrinkage. Similarly, the core insert is also enlarged. However, the compensation is only made to the external profile of the inserts; the dimension of the cup profile is unchanged. The modified files are exported in STL format. After analysis with CAD software (Fig. 9.12) to ensure no STL errors, the files are sent to the RP machine for the fabrication of the tooling parts.

Fig. 9.12 STL display on CAD screen.

Step 2: Fabrication

EOSINT M250 is used to build the mold inserts, based on a powder mixture of bronze, nickel and copper phosphide. After removing the inserts from the EOS machine, epoxy infiltration process is carried out so as to improve the strength and density of thc mold inserts. Concurrently, the mold base can be fabricated in-house or ordered from mold base manufacturers. The fabrication process using RP is shown in Fig. 9.13. The parts have an average hardness of 60 HRV and surface roughness of approximately 4μm.

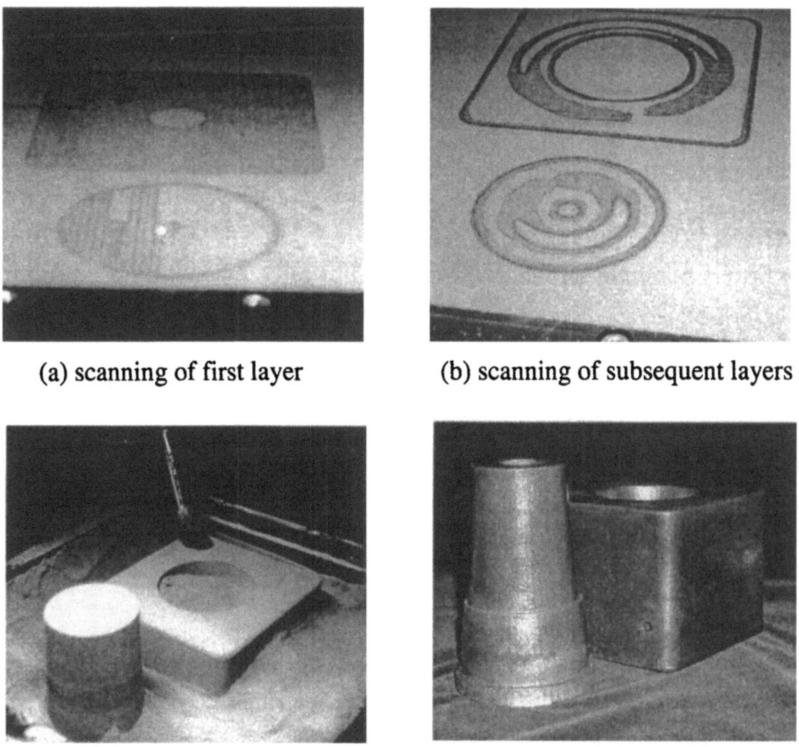

(a) scanning of first layer

(b) scanning of subsequent layers

(c) green parts with excess powder

(d) completed epoxy infiltrated tooling

Fig. 9.13 Fabrication process of cavity and core insert.

Step 3: Polishing and Fitting

Manual polishing process has to be carried out to improve the surface roughness of the core and cavity inserts. At this stage, the mold inserts and the mold base are brought together for assembly. The completed EOS parts are trimmed to the exact dimensions for fitting into the mold base and suitably polished before they are used for injection molding (Fig. 9.14).

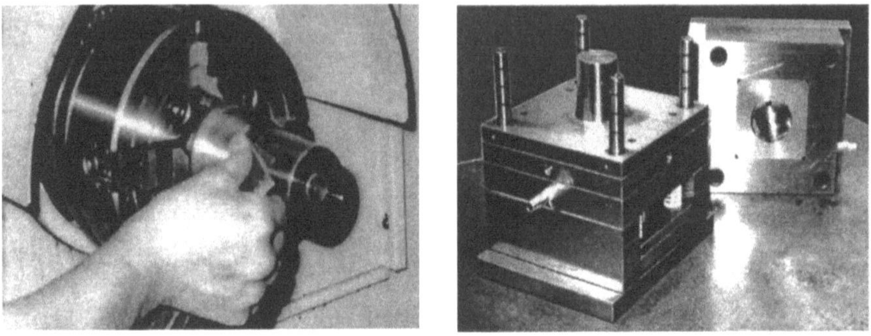

Fig. 9.14 Polishing and assembly of mold.

Step 4: Injection Molding

The assembled mold is put through plastic injection run using three types of thermo-plastic materials with different viscosity. These are SAN, ABS, and PP. After 1,200 moldings (as shown in Fig. 9.15), the core and cavity inserts show no visible signs of wear. Two different rates of cooling are used, one at 40°C and the other one at 60°C. At each temperature, 200 pieces of cups of each material are injected.

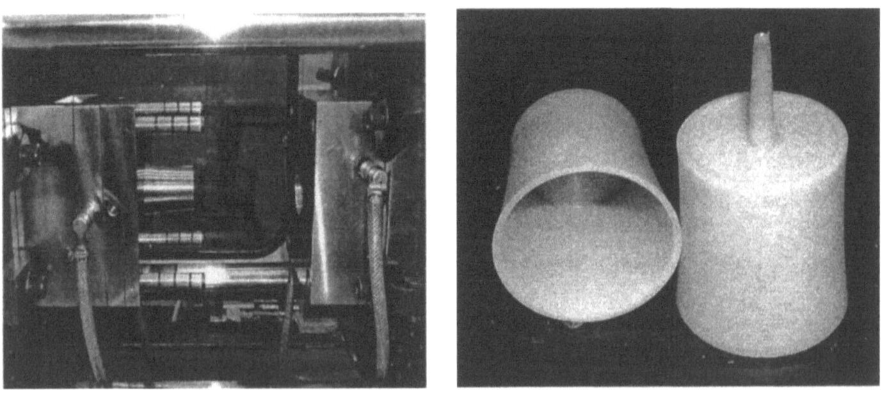

Fig. 9.15 Injection molding and injected cups.

9.3.2 Performance evaluation

The *lead time* to mold parts using the injection mold with EOS inserts is shown in Table 9.2. That molded by conventional processes is given in Table 9.3 for comparison. The lead time of injection mold fabricated by the EOS process is shorter than that fabricated through the conventional process by two days (or 14 % shorter). Although the final cost of the RT-fabricated mold assembly may not always be lower than that by the conventional technique, the main benefit of using RT-fabricated mold, in addition to shorter lead time, is the expected faster subsequent changes. With subsequent manufacture of the core and cavity inserts derived or modified from the first design, the mold can be re-assembled for new injection much cheaper and faster than the conventional tooling approach.

Table 9.2 Process time chart using RT

Process			Day											
			1	2	3	4	5	6	7	8	9	10	11	12
Design	Product Design		■	■										
Design	Mold inserts and Mold base design				■	■								
Fabrication	Mold Inserts	EOS Process					■	■	■	■	■			
Fabrication	Mold Inserts	Inserts Removal									■			
Fabrication	Mold Inserts	Epoxy Infiltration										■		
Fabrication	Mold Inserts	Polishing & Fitting											■	■
Fabrication	Mold base						■	■	■	■	■	■	■	

Table 9.3 Conventional process time chart

Process			Day													
			1	2	3	4	5	6	7	8	9	10	11	12	13	14
Design	Product Design		■	■												
Design	Mold inserts and Mold base design				■	■										
Fabrication	Mold Inserts	Rough Machining					■									
Fabrication	Mold Inserts	Heat Treatment						■								
Fabrication	Mold Inserts	EDM Electrode Preparation							■	■						
Fabrication	Mold Inserts	Precision Machining									■	■				
Fabrication	Mold Inserts	EDM											■	■		
Fabrication	Mold Inserts	Polishing & Fitting													■	■
Fabrication	Mold base						■	■	■	■	■	■	■			

As listed in Table 9.4, there is a large gap in the mechanical performance and durability between EOS material and conventional tooling materials, e.g. P20. The density, hardness and thus durability of RP tools are still not as good as the conventional tool. According to the result of Table 9.5 the average deviation is 0.16mm in *X*-direction, 0.72mm in *Y*-direction and 0.42mm in *Z*-direction. The set of data indicates that shrinkage in the X, Y and Z directions is not the same. The further optimization of the process parameters in EOS machine is thereby needed to improve part accuracy.

Table 9.4 Performance comparison of RP tools vs conventional tools [15]

	Performance			
	Surface finish	Density(%)	Hardness	Life time
EOS Material	4.14~6.03 μm	60-70%	HRV 60	5,000~10,000* shots
Conventional Tooling Material (P20)	0.2~0.4 μm	100%	HRC55-60	1,000,000 shots

(The life time expected is 5,000-10,000 shots.)*

Table 9.5 The result of dimensional measurement (on cavity insert)

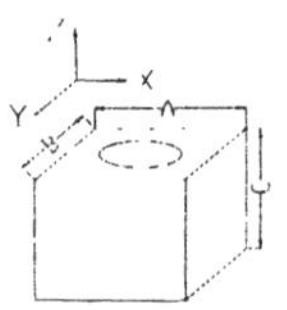	A (X- direction)				B (Y- direction)				C (Z- direction)			
	Measured		Avg	Ideal	Measured		Avg	Ideal	Measured		Avg	Ideal
	1	121.825	121.839	122	1	121.410	121.278	122	1	91.49	91.58	92
	2	121.825			2	121.220			2	91.64		
	3	121.869			3	121.211			3	91.61		
	4	121.837			4	121.272			4	91.75		

In terms of injection cycle time, Table 9.6 compares the cycle time and part quality in different cooling conditions for the three types of thermo-plastic materials with different viscosity (PP, ABS and SAN). It can be seen that the application of conformal cooling channel can greatly reduce the cycle time by 30% and diminish part shrinkage. Although the overall mechanical properties of RP tools are still not as good as the functional tooling manufactured conventionally, RP-produced tools as a prototype tooling, are acceptable in addition to lead time and cost saving.

Table 9.6 Injection parameter, cycle time and parts quality

Cooling System	40°C Mold Temperature			60°C Mold Temperature			Without Conformal Cooling Sys		
Material	PP	ABS	SAN	PP	ABS	SAN	PP	ABS	SAN
Injection Pressure (bar)	40	55	74	40	55	74	40	55	74
Injection Temperature(°C)	270	230	250	270	230	250	270	230	250
Cycle time (Sec)	36	34	35	40	34	42	50	48	52
Parts surface	Rough	Smooth	Smooth	Rough	Smooth	Smooth	Rough	Smooth	Smooth
Parts Shrinkage Opposite sprue	Shallow Dent	Shallow Dent	Shallow Dent	Shallow Dent	Shallow Dent	Shallow Dent	Deep Dent	Deep Dent	Deep Dent

9.4 Application to EDM electrode fabrication

Electrical discharge machining (EDM) is a non-conventional machining process (Fig. 9.16) where material is removed electro-thermally by a series of successive discrete discharges or sparks between two electrically conductive objects, i.e., the electrode and machined work-piece. It is very commonly used in die and mold machining on which certain features cannot be machined using the traditional CNC machines. The performance of the EDM process, to a large extent, is dependent on the material and design of the electrodes. EDM electrode material must have the basic properties such as electrical and thermal conductivity, a high melting temperature, low wear rate, and resistance to deformation during machining.

The electrodes for EDM are normally fabricated using the conventional machining methods. The cost of manufacturing electrodes is generally determined by the complexity of the geometry and accuracy required. Uniformly-dimensioned electrodes are cheaper to fabricate. For complex three-dimensional shapes, the electrodes may be fabricated by machining, casting, electro-forming or metal spraying. Fabrication of EDM electrodes using RP technology [18-20] can reduce the cost of fabricating electrodes with complex geometry. RP potentially provides a method of manufacturing electrodes quickly with minimum manual intervention and offers great potential for reducing lead time and tooling costs. Furthermore, it can eliminate lengthy conventional machining process for complex electrodes and at the same time, the electrode material that is difficult to machine can also be used in the RP system for direct electrode fabrication.

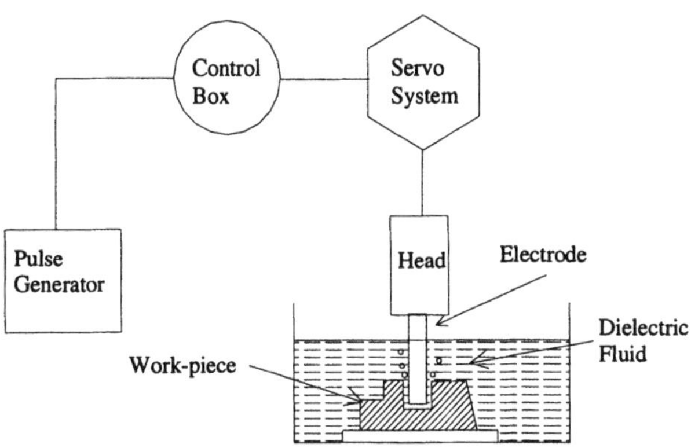

Fig. 9.16 EDM die sinking process.

9.4.1 Indirect sintered or formed electrodes

Sintered EDM Tools

The commonly used materials for EDM electrodes are graphite, copper, copper alloys, copper-tungsten, brass, silver-tungsten, and steel. Each material has its pros and cons on the machining performance. Materials having good electrical and thermal conductivity with a high melting point are preferred for electrode fabrication. Before direct fabrication of electrode using RP systems, furnace sintering can be used to determine the proper composition of constituents suitable for layered manufacturing. A few potential electrode materials have been identified [18,21,22] that have comparable EDM performance as commercial ones. The powder systems used to make these electrodes include zircon boride with copper (ZrB_2/Cu), titanium carbide with nickel (TiC/Ni), copper tungsten (Cu-W), etc. It has also been found that ZrB_2/Cu electrode has a wear rate between copper and graphite, but the material removal rate is low, and the relative electrode wear rate is higher than conventional electrodes. It shows a good potential to be applied in forming complex electrodes using SLS [21].

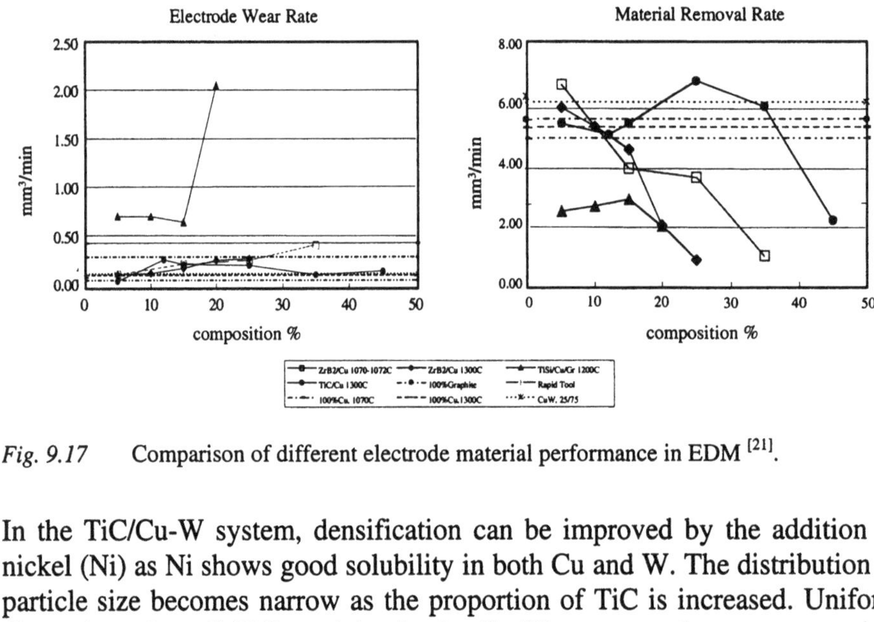

Fig. 9.17 Comparison of different electrode material performance in EDM [21].

In the TiC/Cu-W system, densification can be improved by the addition of nickel (Ni) as Ni shows good solubility in both Cu and W. The distribution of particle size becomes narrow as the proportion of TiC is increased. Uniform dispersion of small TiC particles in the Cu-W system and a narrow particle size distribution provide the possibility of obtaining dense electrodes. With increasing TiC the relative density first increased and then decreased, whereas the electrical resistivity first decreased and then increased. EDM electrodes with addition of TiC show good performance in surface finish. This is an important characteristic as RP-sintered EDM electrodes are expected to be used as finishing electrodes. The surface roughness of most specimens is less than that machined using commercial electrodes. Electrodes with 15% TiC show the highest relative density, lowest electrical resistivity, and good EDM performance, i.e. lowest tool wear ratio (TWR) and highest material removal rate (MRR) at low current, and the best surface finish not only at low current, but also at high current [22].

Electro-formed Tools

Lee, Guesoz *et al.* [23] developed a rapid tool manufacturing system based on LL and thermal spraying method. The integration of rapid prototyping and electroforming for tooling applications has been reported by Yang *et al.* [24]. The feasibility of fabricating EDM electrodes with RP and electroforming process was reported by Prasad *et al.* [25]. Arthur *et al.* [20] investigated the use of electroplating LL patterns to build EDM electrodes to a thickness of 180mm, as shown in Fig. 9.18. They also conducted the work on copper electroformed electrodes from LL cavities. However, the potential for use is found mainly in roughing and semi-roughing operations with efficient machining rates and the separation of the electroform from the LL pattern is a problem. Prasad *et al.* [25] therefore used an electro-formed shell of copper, backed with a suitable material, as an EDM electrode. Their study concluded that the electroformed copper electrode seems to possess a potential for use as EDM tool. Dover and others [26,27] used electro-deposition method to manufacture metallic parts. A copper and nickel sulphate system was used to produce press tools. In summary, electroformed copper electrodes do have good potential for use as EDM tools especially for complex geometry. However, the electro-plating cannot sustain long in EDM and the tool can be easily deformed during rough machining and thus semi-finishing to finishing operations are more appropriate in using these tools. An alternative is therefore to use the direct laser-sintered metal tooling.

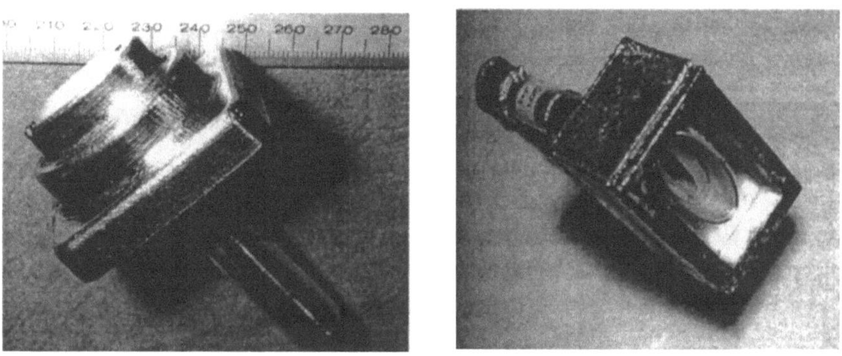

Fig. 9.18 SLA Electroplated core and cavity EDM tool [10, 20].

9.4.2 Direct laser-sintered electrodes

Rapid tooling of injection molding and EDM electrodes by selective laser sintering is probably the most successful tooling applications based on RP techniques. The commercialized RapidTool® from DTM Corporation and DirectTool ™ from EOS GmbH are mainly applied to injection molding, and have achieved promising results for industry use. There are still few successful cases reported on producing EDM tooling using selective laser sintering (SLS) or direct metal laser sintering (DMLS) so far. The main problems are high porosity, quick wear and surface roughness of the fabricated tools. Stucker [18] and Stucker *et al.* [19] were the first to mention EDM electrode making based on SLS. A Cu-ZrB_2 powder mixture has been used for sintering the electrode. At the high sintering temperature, the sintered electrodes have shown comparative EDM performance as commercial ones in certain areas. Dürr et al [28] reported the rapid tooling of EDM electrodes by means of direct laser sintering based on bronze-nickel and copper phosphite powder mixture and EOSINT M 160 machine. It is found that the influencing factors affecting EDM performance include: process parameters (i.e. laser power, hatch distance and scan rate), subsequent post treatment, powder characteristics, source of laser (CO_2 vs. Nd/YAG), and operating environment. EDM performance on material removal rate and tool wear rate relative to various porosities and process parameters have been studied.

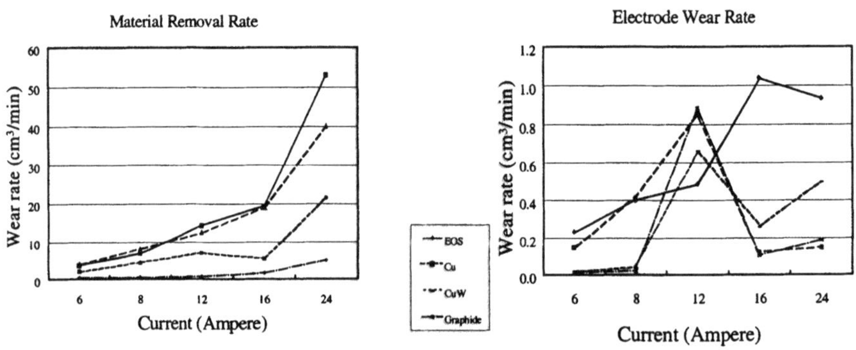

Fig. 9.19 EDM machining performance on steel work-piece.

A detailed EDM performance evaluation on both EOS electrodes and EOS work-pieces can be found in Reference [29]. Fig. 9.19 shows the material removal rate and tool wear rate of EOS electrodes on a steel work-piece at different current, in comparison with other commercial electrode materials. It concludes the previous finding [21] that CuW-steel combination gives good results due to a significant reduction in machining time, good machining rate and relatively lower electrode wear rate. The wear rate is the same for both commercial and EOS electrodes. Roughness of the machined surface varies with current. Machining time is shortened when the applied current is decreased. In general, the performance of EOS electrodes is poorer than the commercial ones, and shape deformation of EOS electrode is much faster than the commercial ones. Infiltrated EOS parts are however not suitable for use as electrodes because of the low-melting point component (epoxy), as evident by unstable sparking formation and difficulty to control the process. Good tool wear rate and material removal rate will therefore not be expected [29]. New material systems need to be developed and further researched to improve the mechanical strength and to reduce the porosity of RP-fabricated electrodes.

9.5 Summary

Rapid prototyping is increasingly applied to rapid tooling development that covers applications to injection molds, EDM electrodes, press tools, metal-based tooling, etc. With the aid of rapid prototyping techniques, rapid tooling is able to produce reasonably cost-effective prototype tools. The application to injection mold, in particular, cavity and core inserts, has proven to be workable to inject plastic parts in the mid-volume range. Recent advances in material study and development have made it possible to fabricate EDM electrodes with complex geometry for die-sinking process. Direct tooling will certainly have more potential than indirect tooling due to the capability of manufacturing functional and wear-resistant tools. But, that will require further research to be carried out to improve the quality of the tools in comparison with the conventional tooling. The RT techniques also suffer from a lack of basic understanding on the fundamentals. Modeling and numerical simulation should provide better comprehension of the energy absorption phenomena in laser-based methods, wetting effects, chemical reactions, etc. and hence the RT process itself [30].

9.6 References

1. P.D. Hilton, P.F. Jacobs, *Rapid Tooling: Technologies and Industrial Applications*, Marcel Dekker, Inc, New York, 2000.
2. A. Rosochowski, A. Matuszak, *J. Mater. Proc. Tech.*, Vol.106 (2000), 91.
3. C. Ainsley, H.Q. Gong, *RP. J.*, Vol.5, (1999), 35.
4. DTM Corporation, www.dtm-corp/applications, 2001.
5. EOS GmbH, Germany, www.eos-gmbh.de, 2001.
6. Sandia National Laboratory, USA, www.sandia.gov/LabNews/LN12-05-97/lens_story.html, 1997.
7. SDM, Stanford University Rapid Prototyping Laboratory, www-rpl.stanford/planner.html, 1997.
8. F. Klocke and U. Clemens, *Proc. of the 5th Eur. Confer. on RP. and manuf.*, (1996), 211.
9 CMB, Controlled Metal Build-up, Fraunhofer Insititue of Prodution Technology, Aachen, Genmany, 1999.
10. R.C. Soar, A. Arthur and P.M.Dickens , *Proc. of the 2nd Inter. Confer. on RP. and Tooling Res.*, edited by G. Bennett (1997), 65.
11. 3D Keltool, 3D Systems, Inc., Valencia, CA, www.3dsystems.com, 2001.
12. P. Jacobs, Proc of *the 2nd National Confer. on Development in RP. and Tooling*, (1997).
13. A.D. Venus, S J van de Crommert, S. O'Hagen, *Proc of the 2nd National Confer. on Development in RP. and Tooling*, (1997).
14. Y.H. Tan, BEng Thesis, National University of Singapore, (1999), 27.

15. X.H. Wang, J.Y.H. Fuh, Y.S. Wong, *Proc of Inter. Confer. of Precision Engineering, Singapore*, (2000), 455.
16. IMOLD^{TM}, ManuSoft Plastic Pte. Ltd, Singapore, www.manusoft.com.sg, 2001.
17. E. Sachs, S. Allen, M. Cima, E. Wylonis, and H. Guo, *Proc. of the 7th Solid Freeform Fabrication Symp.*, Austin, USA, (1996), 448.
18. B. Strucker, *Proc of Solid Freeform Fabrication Symp.*, (1995), 278.
19. B.E. Stucker, W.L. Bradley, S. Noraset-thekul, P.T. Eubank, *Technical Report*, Texas A&M University, 1996.
20. A. Arthur, P.M. Dickens, R.C. Cob, *RP. J.*, vol.2(1), (1996), 4.
21. H.M. Zaw, J.Y.H. Fuh, A.Y.C. Nee, L. Lu, *J. Mater. Proc. Techn.*, vol.89-90 (1999), 182.
22. L. Li, Y.S. Wong, J.Y.H. Fuh, L. Lu, *Mater. Design*, (2001), in print.
23. W.E. Lee, E. Gursoz, F.B. Prinz, *et al.*, *Manuf. Rev.*, vol.3(1), (1990), 40.
24. B. Yang, M.C. Leu, *CIRP Annals – Manuf. Tech.*, vol. 48(1) (1999), 119.
25. K.D.V. Prasad, P. Christodoulou, V.S. Subramanian, *J. Mater. Proc. Tech.*, vol.89-90 (1999), 231.
26. S.J. Dover, C.E. Bocking, G. Bennett, *Proc. of 1st National Confer. on RP. and Tooling*, Mech Eng Publications Ltd, (1995), 157.
27 S.J. Dover, C.E. Bocking, *RP. News Letter*, Iss. 9 (1996), 157.
28. *H. Dürr, R. Pilz, N.S. Eleser, Computers in Industry*, Vol.39 (1999), 35.
29. M.W. Khaing, *MEng Thesis*, National University of Singapore, (2001), 95.
30. N.P. Karapatis, J.-P.S. van Griethuysen, R. Glardon, *RP J.*, Vol.4 (2), (1998), 77.

INDEX

MIX
Papier aus verantwortungsvollen Quellen
Paper from responsible sources
FSC® C105338

If you have any concerns about our products,
you can contact us on
ProductSafety@springernature.com

In case Publisher is established outside the EU,
the EU authorized representative is:
Springer Nature Customer Service Center GmbH
Europaplatz 3, 69115 Heidelberg, Germany

Printed by Libri Plureos GmbH
in Hamburg, Germany